What people are saying about

Other Paradises

In *Other Paradises,* Jessica Sequeira reads generously, cleverly, with clarity and elan. She is spot on about how technology is recreating literature irl, but even better, Sequeira's essays recreate criticism, imaginative response by imaginative response. Reading this book one imagines yet another paradise, one where we don't read any critic who can't match Sequeira for energy and acuity.
Daniel Bosch, Poet and translator, lecturer in English literature at Emory University, senior editor of Berfrois

A rich, testing and pleasing book... I'm still thinking about its discussion of our relationship with technology: its endless variety, the surprise of how playful it can be, and the joyous inventiveness of the human mind capable of writing like this.
Jon Lindsay Miles, Translator of Haroldo Conti's Southeaster, publisher at Immigrant Press

Other Paradises

Poetic Approaches to Thinking in a Technological Age

Other Paradises

Poetic Approaches to Thinking in a Technological Age

Jessica Sequeira

Winchester, UK
Washington, USA

First published by Zero Books, 2018
Zero Books is an imprint of John Hunt Publishing Ltd., Laurel House, Station Approach,
Alresford, Hants, SO24 9JH, UK
office1@jhpbooks.net
www.johnhuntpublishing.com
www.zero-books.net

For distributor details and how to order please visit the 'Ordering' section on our website.

Text copyright: Jessica Sequeira 2017

ISBN: 978 1 78535 585 1
978 1 78535 586 8 (ebook)
Library of Congress Control Number: 2017933413

All rights reserved. Except for brief quotations in critical articles or reviews, no part of this book may be reproduced in any manner without prior written permission from the publishers.

The rights of Jessica Sequeira as author have been asserted in accordance with the Copyright, Designs and Patents Act 1988.

A CIP catalogue record for this book is available from the British Library.

Design: Stuart Davies

Printed and bound by CPI Group (UK) Ltd, Croydon, CR0 4YY, UK

We operate a distinctive and ethical publishing philosophy in all areas of our business, from our global network of authors to production and worldwide distribution.

Contents

To all the ludic spirits — keep close.

About This Book

What is an imaginative response to technology?

Something that does not reject technology,
but draws on its ways of thinking for non-productive ends.

Something that self-consciously transforms itself and its processes into a subject.

Something that makes process into play, possibly useful but also diverting.

Imaginative responses to technology can be the basis for whimsies, utopias and visions that build castles from fact and fiction.

They can engage objects, places and theories of knowledge to transcend the limitations of geography and time.

And they can go hand in glove with literary modernism.

In each chapter, a response to some aspect of technology is elaborated, then twisted into new shape.

Acknowledgments

Versions of these essays were published by *Berfrois, Drunken Boat, Entropy, Gauss PDF, Queen Mob's Teahouse, The Missing Slate* and *3:AM*.

My thanks to the editors of those publications, for being generous with the needed encouragement, and expanding my ideas of poetry and play.

Preface

Growing up in California, a faith in technology was just a part of the landscape, something like redwood trees or swimming pools. That unquestioned faith was in the corporate buildings lining the roads, the Internet cafés, the paintings of David Hockney. It was so much there that it was hard to know how to talk about it. 'I know what you mean,' a friend said when I told him I was writing these essays, 'but you can't possibly take on such a huge theme. Besides, I thought you liked to hang out with poets.'

Valid objections, to which I say: this book is about imaginative *responses*. Whether or not one chooses to embrace new gadgets with open arms, whether or not one lives in the heart of Silicon Valley, technology and people's responses to it have an effect. Perhaps many responses don't even think of themselves as responses, in which case reframing them is useful, giving them new meaning by placing them in new context. Any 'imaginative response to technology' has to be epistemological as well, a change in ways of understanding or thinking. *Other Paradises* is about imaginative responses to science and technology in cities ranging from Tokyo to Mexico City, Oruro to Calcutta. It is about the ways that technology is understood and engaged with in different global contexts, and about how people build 'other paradises' fully conscious of the alternative they offer to the dominant paradigm of technological progress. It's not necessary (and perhaps impossible) to be 'against technology', but one can draw on its potential in whimsical, inventive ways that transcend use value. Many possible methods exist of defining modernity and the relationship between play and knowledge. Why do people deliberately choose to play with ideas considered antiquated? Why do they elect to act in non-productive ways, when they have the option to do otherwise? Or maybe I can ask the question in reverse: What comes to mind when we think of technology? That which is practical, efficient, light, invisible, tiny, fast, optimistic, new, constantly updated, planned and determined. So how can we explain the search for its opposite, that which is useless, inefficient, heavy, physically present, big, slow, dystopian, old, obsolete and governed by chance? The question of what motivates the search for 'antiquated' forms strikes deep into the heart of value. Are people simply following trends? Are they idiots? Are they sentimental? Are they artists? Are they interested in kitsch? Are they uninformed? Are they poets? The question of what drives participation in 'unproductive' behavior, opposed to the technological ideas that supposedly define our modern age, or that divert these ideas down strange new paths, is fascinating. It's connected to ideas of freedom, creation of the self, self in relation to others, consciously chosen communities, utopias that do not use technology or use it in non-productive ways, apocalypse, dystopia, autobiography and history. People are often very aware that the worlds they're creating are based on 'false premises' and are not on the cutting edge, that they are seen as either far behind or too early for their times. Yet they embrace these worlds regardless, creating artificial paradises, poetic experiences or alternative means of forming connections, which may be masks or simulacra but are no less real for it.

Creative responses, defined as 'art' or not, exist all over the world, taking forms such as Maker culture in Istanbul that repurposes existing technology, magic in nineteenth-century France and retrotech in Tokyo that intentionally favors obsolete technologies such as the fax machine. The attitude taken toward technology by these projects is curious but whimsical, open to new developments independently of whether or not the ideas behind them produce profit. Very different values may drive such behavior, but the effect of such fantasies is to sharpen the intensity with which we perceive both mediated reality and the non-human world.

The opposite of technology's focus on ideas with application is not the rejection of technology, but the embrace of wondrous, sensuous ideas with a sense of play. 'Technology' as a shifting problem and 'imaginative' responses are nothing new. Perhaps this selection is eclectic, but I wanted to try to open up ways of thinking, my own to start with, rather than close off any paths. Poetic responses to technology are not afraid of technology, and can discover in scientific structures of thought great inspiration. But they also know that the world does not end there. Beyond or around or through all the simulacra gleams the utopia toward which technology aspires — a reality that is comic and beautiful and true.

Ghost in the Fax

haunting new technologies with old ones as retrotech / Japan

Dear reader, here we are now, you and I. Ghosts, half here, half not. If I reach out and try to place my hand on your shoulder, I won't feel a thing. But I know you're close, so trust me. Follow me into this office, in a place far away. The first thing you'll see is a plaque with Kenji characters in alarmingly bright shades of orange, pink and green, instructing employees on how to receive transmissions. If someone miles across the city presses the right series of buttons, a sheaf of papers will emerge on this side. If the same series is pressed here, the pages will be sent there, in a mirror of the process.

At rest for the moment, the machine calmly hums, anticipating activity. A few muttered complaints can be heard. Better ways exist of going about things, people say. The big box takes up space that might be occupied by a water cooler or Seibo vending machine. But when the admin asks if it's time to swap out the machine, there are protests. The fax, gunslinging hero of electronic devices, resists any attempt at a coup. Nor do his subjects wish to see him go. Despite bureaucratic and logical objections, a barrier of allegiance remains. Big, solid, ferocious, the fax with his beeps and whistles seems far more accessible even in his inefficiency than the inkjet printer that rests in ominous silence, the fiend that jams without warning. 'Alright,' says the admin with a sigh, secretly pleased at the protests. 'He can stay another year.' The fax digs in his heels, Lone Star Ranger defending his ghost town.

Multicoloured neon paradise, glittering city of the future, Tokyo is also an epicentre for the intentional use of obsolete technologies. The fax is beloved, so much so that government reports lament widespread irrational attachment to the technology. Its electronic tones may be endearing, but surely it's not the most efficient way to transmit information? Yet, Tokyo collectors and office workers self-consciously embrace the démodé. Faxes are kept alongside modern computers, and the intoxicating scent of ink and paper perfumes the air. The old technology is present elsewhere as well, in a more complex, hidden way. Conceptually, the fax is nested inside the idea of modern email, and it's not alone. Thousands of widgets, gadgets and toggles are kept tucked away in drawers, anachronistic but unforgotten, haunting the present.

While floating through the office we pass a girl wearing a Hello Kitty T-shirt. Perhaps the obsession with retrotech has to do with Japan's kawaii culture, the obsession with cuteness and fetishism of childishness and nostalgia in the form of manga comics, schoolgirls and plush toys. It isn't necessarily part of the national psyche, and 'national psyche' is a myth, anyway. But the elements have been encouraged so heavily in the entertainment and marketing industries that it may as well be. Kawaii Taishi, ambassadors of cuteness, even travel around the world as representatives of the nation. Tempting as it may be to attribute the prevalence of retrotech to the kitsch charm of machines, however, there's something else at work.

Let us drift away now to see other marvels, and try to draw the veil from these mysteries. Leaving the city, we reach an open field, edged in on one side by a dreaming forest. Of course, forests can dream; how else do you explain their clearings, their dappled light and shadow? And there they are in the field, hundreds of them. Balls of blue fire, dancing will-o'-the-wisps. These are the *hitodama,* ghosts described by Lafcadio Hearn, a writer born on the Ionian islands of Greece. After a colourful life in the United States and French West Indies, Hearn came to Japan to work as a newspaper correspondent and schoolteacher. His impressions of the landscape and eerie tales of the country take up several books, and *Glimpses of Unfamiliar Japan,* his first chronicle, is full of the ghosts he sees everywhere in nature. 'The very clouds are not clouds, but only dreams of clouds, so filmy they are,' he writes, 'ghosts of clouds, diaphanous spectres, illusions.' He wrote several books of *kwaidan*, performed Japanese folk stories: 怪 (kai) means *'strange, mysterious, rare or bewitching apparition'* and 談 (dan) means *'recited narrative'*.

He had much to tell, for there are many different kinds of ghosts in the culture, not just balls of fire that float in the night. You and I can go and meet some of our first cousins, if you like. In the Shinto religion, the kami are worshipped spirits, and include not just the dead, but the qualities possessed by those spirits. There are ikiryō (also called shōryō, seirei, ikisudama), which leave the bodies of living people to haunt others across distances. There are shiryō, which do the same but emerge from the spirits of the deceased. And there are yūrei, which roam the city, restless and seeking revenge; these appear in the tale of *Banchō Sarayashiki,* in which a servant girl is thrown down a well by her master after she refuses to give him love, and he accuses her of stealing a Delft plate from the family. Ghosts are everywhere, busy laughing, crying, loving, plotting, dancing and sleeping just like humans. Whether they lie in tranquility awaiting visits, or whirl about buildings, they remain a link to the past.

But if Japanese tales allow inanimate objects like elements to have ghosts, isn't it possible that obsolete machines have them, too? Technology left behind takes on a phantom presence, both here and not here, physically present yet not considered useful. It isn't simply nostalgia, but a reminder the 'now' hasn't always existed. Incarnations of the ghost-of-technology-past inform and shape the ghost-of-technology-present, in turn creating the conditions for the ghost-of-technology-future. The fax machine at the Tokyo office rests peacefully, but does not rest in peace. Dormant, it is still prepared for a shoot-off, a battle for the high stakes of whether or not it will be forgotten.

Oblivion: such a beautiful word, so terrible when it occurs.

When a technology appears, first there is change; then, forgetting. One technology replaces its predecessor with awful rapidity, annihilating its memory. It may be possible to send and save as much as one likes, in an almost infinite archive, but the mediums that perform this task are jettisoned and replaced with astonishing speed. Not all is lost, however. Physical reminders keep the past alive. Museums full of objects in glass cases, on the one hand; but also old technologies and objects outside the institution, half dead and half alive. The sight of them links what was, and what is coming into being. Progress is not simply linear, and does not forget what came before it. Each step adds something different, a new input, yet the previous value hovers for a time 'embedded' in the present. Amnesia is the natural course; the challenge is to ensure the opposite.

In Japanese culture, oblivion is a special concern. Moderns suffer from 'character amnesia', and forget how to write kanji characters as they are so used to phonetic transcription. Oblivion threatens entire sign systems. During the Edo period, a branch of Japanese mathematics was developed in parallel with Western math, a product of the country's global isolation at the time. Japanese geometrical problems and theorems were written on wooden tablets, and placed as offerings at Shinto shrines or Buddhist temples. In Floating World Era Japan, sacred mathematics was produced and taught in local schools by samurai, who 'had originally been independent warriors but settled down in the seventeenth-century to become a local aristocracy of well-educated officials and administrators. It was the samurai class that supplied mathematicians to create the sangaku and work out problems,' as Fukagawa Hidetoshi and Tony Rothman write in *Sacred Mathematics: Japanese Temple Geometry*. The interested reader (anyone who likes cryptics, crosswords and circular puzzles; sound familiar, my fellow ghost?) may consult the book *Japanese temple geometry problems: Sangaku*, published by the Charles Babbage Research Centre in

Winnipeg, Canada.

A brief digression, which may prove not to be so brief, or so much a digression: Who is this Babbage, of the Charles Babbage Research Centre? He was the inventor of the difference engine, a mechanical contraption that received input to produce output, essentially a calculator. His machine worked on the basis of functional equations, in which the result forms part of the input for the next round, one abstract nested in another. The machine was able to solve equations based on addition and subtraction. Later, Babbage improved it with the analytical engine, which could also take on multiplication. His friend Ada Lovelace compared the machine to a Jacquard loom, since just as with a shuttle, the old does not disappear, but helps to determine the next starting point.

For an object to persist even after it has been rendered obsolete or died, finding a way to be physically present rather than existing solely as data, is a victory in itself. (Back in Tokyo, the fax at the office fires off two ebullient sequences.) Neither do humans simply disappear, any more than data does. Their memories can also hover as ghosts. Those famous or loved enough might get a plaque, or a tombstone, or a spare thought now and then from the living.

On the side of Walworth Clinic in London, Babbage's blue plaque rests content, reading: '1791 to 1871. Mathematical genius, astronomer, inventor and "Father of Computing".' Babbage also appears stuck on thousands of envelopes in the form of a popular postage stamp, sporting a scarlet coat and bow tie, his bright yellow head filled with an inverted pyramid of numbers. Multiple versions of the difference engine continue to exist as physical objects at the Science Museum in London. Whether or not Babbage is now a ball of fire floating through the night, his inventions have not disappeared, but persist into the present in phantom form, the basis for what came next: the modern computer.

While at Cambridge, Babbage was a founding member of the Ghost Club, which 'collected evidence and entered into a considerable correspondence upon the subject'. How does one ensure that ghosts survive, that traces are left after the death of the physical body? Either through a physical monument, or through writing. As Babbage writes in the preface to his *Passages in the Life of a Philosopher*:

> Some men write their lives to save themselves from ennui, careless of the amount they inflict on their readers. Others write their personal history, lest some kind friend survive them, and, in showing off his own talent, unwittingly show them up. Others, again, write their own life from a different motive — from fear that the vampires of literature might make it their prey... This volume does not aspire to the name of an autobiography. It relates a variety of isolated circumstances in which I have taken part.

One of these 'isolated circumstances' was recently uncovered on a tapestry. Shortly after Lafcadio Hearn arrived in the town of Matsue, he was shown a series of *kakemono*, or hanging scrolls. He described one this way:

> In the upper part of the painting is a scene from the Shaba, the world of men which we are wont to call the Real — a cemetery with trees in blossom, and mourners kneeling before tombs. All under the soft blue light of Japanese day. Underneath is the world of ghosts. Down through the earth-crust souls are descending. Here they are flitting all white through inky darknesses; here farther on, through weird twilight, they are wading the flood of the phantom River of the Three Roads, Sanzu-no-Kawa.

What he doesn't tell is how he walked on, and saw another

kakemono, with its image mounted on a silk panel. What was painted there, in exquisite jewel-like detail, was a difference engine, carefully sewn in black-and-gold silk. It looked dormant, but in reality was tabulating functions based on polynomial coefficients, using ink and thread count as its initial input. Are you surprised by what I am telling you? You don't look taken aback, nor should you be, for it makes sense. Babbage began to create his first difference engine in 1821, and his analytical engine in 1834. On July 8, 1853, Commodore Matthew Perry of the United States Navy sailed his frigate USS *Susquehanna* into Tokyo harbor. Japan opened to the West, and knowledge of European inventions entered the country.

A Japanese artist chose to dedicate one of his works to a recent invention, not from the Edo period of Japanese sacred geometries and temple mathematics, but from the new Meiji age, in which the West was no longer considered a demon. In this way, the calculator of the difference engine appeared in his work. Using ink and thread, this third difference engine, one never dreamed by Babbage, calculated our ghosts, and this story. Do you still believe me? After seeing such strange, eerie things, you must have asked yourself who we really are. Ghosts, yes, but what kind? Just know that in a world in which ghosts coexist with the living, the marvellous is always possible. We are here, we are not here, binary, infinite.

Story of a Drifter

> I was just a floater down those Tokyo streets. Surrounded by bright lights, I would have been so lost, if it hadn't been for the Boss. To most of the world he was just one of those yakuza bad guys. 'Hey, get over here,' he said that morning. Broad daylight and a knife in my back. It was pointy, to state the obvious. 'You can do something for us,' he said. 'You're going to help round them up.' He took a long drag from his

cigarette. The others laughed (that's when I noticed there were others) and showed their big donkey teeth.

All of it was just something I fell into. I never meant to be a yakuza man. It's still weird. Maybe later, when there's time to reflect, everything will seem natural, new life spilling from old. Organs from the belly of a dead man. No, I don't like that image at all. Man sprawled in a street after being popped off; it's not something I've ever seen in real life, anyway. Except in movies about yakuza men.

In general I'm a peaceful guy. Even after I joined, nothing changed about my routine. Strong morning coffee. Yellow suit (it defines me). Nice shoes. Sunglasses. Hank Mobley's *Soul Station* on vinyl, repeat. You wouldn't believe how much I love that album. I've had to replace it three times already. Lookin' good, I say to myself in the mirror, do the finger-gun. Then I'm out the door, down the stairs and into the Kabukicho district. What a neighborhood.

Like I was saying, I'm just a normal guy. I even work in a photocopy shop, zero glamour. What, photocopies here? You'd be surprised by how many people want things Xeroxed. Paperwork for a Visa, housing this and that, legal-type stuff. Sometimes girls come in asking for copies of books, Banana Yoshimoto and that sort of thing, so when it's necessary they can pull out lines about moonlight and shadows and cherry blossoms.

I was just making my way to the shop when I felt it, the knife in my back. The Boss explained very clearly what we were going to do. Collect clunky old machines that people wanted to throw out. 'And then?' I asked. He grunted. 'We have to round them up. Some of 'em we take apart and repurpose, the ones no good for anything else. Most we try to relocate. Find 'em good homes.'

'Sounds like they're the ones running the show here,' I grumbled. 'In a way they are,' he said, unfazed. 'People think we're a gang, and I guess technically we are. But we're protectors of those old machines. They're the real warriors. The samurai.'

Getting dragged into that job hadn't fazed me. It was surprisingly easy to get used to, and better than the copy shop, that's for sure. But I was surprised by what he said. And the really big surprise came later, when I found out that on top of everything else, I was a ghost. We all were. The pointy knife had killed me. Legends were already springing up around the city about a yakuza ghost gang. And I ... was one of them. It wasn't so bad though, as long as I could keep busy. And my yellow suit was just as bright in the afterlife.

Ada Lovelace, she was the only one I missed. That's what she called herself, the girl I went to see sometimes. Bangs and short dark hair and big eyes, pale skin that glowed, lips like a cherry ice. She pouted and looked so fragile, pouring out tea in her soft lavender sweater. But don't be fooled. I told the Boss about her once, in one of those virile moments at the bar, sharing things over sake. (Ghosts drink, too.) Went on and on about how feisty she was. How I missed her. Got a little weepy, maybe.

Next day she was there with us. That pointy knife of the Boss's must have made her a ghost, too.

How all this got started beats me. I don't ask questions, just take the machines to the warehouse to get broken up or relocated. Sometimes, in the middle of a run, I stop at the copy shop. Maybe I'm sentimental; maybe I think the past never really disappears in the present. I go and run off a stack of Banana Yoshimoto's, for old time's sake. Without fail, some pretty girl with pink-and-orange barrettes and a schoolgirl skirt comes by a few minutes later, giggles and picks one up off the counter. Sometimes, but rarely, she shows up later in our gang.

Optic Nervous

juxtaposing rapid content and slow arrangement / Bay Area

For a long time, when I visited Silicon Valley, I thought of it as the locational equivalent of Musil's man without qualities. A 'not yet' place that rejects a fixed definition and essence, a place defined by what it does rather than what it is. Being from this place doesn't seem to mean too much, or least doesn't give one a solid sense of identity, as being from *fin-de-siècle* Paris might. Most people here are recent immigrants anyway, from some other nook of the world. This might be the condition of all places in the future, but here and now, it simply feels a bit like a white stage backdrop awaiting events — like being from nowhere, with access to everything but in essence nothing, the unbearable lightness of bitmap.

If this were true, the historian's or critic's desire to provide context might seem out of place, as if it were missing the point. Whatever happens already seems halfway inside the Internet, or the mind of some engineer, an abstraction that doesn't benefit from cultural analysis. An illusion, perhaps, but a convincing one, which the 'post-history' of technology encourages. Yet what if this very non-placeness, this non-humanness, could become the basis for a different literary approach? Lots of people here complain that the software is in charge. What if it began to write the books, too?

These were my thoughts as I picked up *Registration Caspar*, published by Ugly Duckling Presse. It was written by J. Gordon Faylor, managing editor of SFMOMA's *Open Space* and editor of the online publisher Gauss PDF. Faylor, a sort of Bay Area Beckett, gives his book a basic plot, but this is only important at a secondary level, as its real innovations occur at the level of procedure. Via email, Faylor told me that many of his sentences developed gradually and endogenously (expanding from within), emerging from a practice central to his work in recent years, the palimpsestic treatment of spambot texts. He would find some hideously garbled and clearly automated spam or robotext, then try to 'rewrite' it from scratch, changing most of the language and, more importantly, trying to structure it like a narrative, to lend it the aura of coherent prose.

Sometimes, these texts would integrate chunks of other novels in order to throw off Google's anti-spam algorithms, which abetted the process all the more. But the approach wasn't strictly applied or procedural, and Faylor made many adjustments and additions via his 'own' writing and life experience along the way, whether in the interest of rhythm, argument, conveyance, grammatical obfuscation or deictic play. A lot of the language also came from coding terms and discourse ('chaining', 'execution').

What all this means in practice is that while the structure of a sentence or paragraph might suggest a meaning, the actual words resist comprehensibility. Play of the eye and play of the word need not be foes, and some of the most interesting work comes from integrating them. Here, however, wordplay handily annihilates the visual. Extreme compression marks the style, a pure string of randomly generated phrases united by a falsely informal conversational tone, like someone chattering away to you in a language you don't know with a smile. When dialogue appears, it's stagy, full of kitsch expressions like 'superstar', uttered by people with names that are either absurd or variations on Bob. Outré adjectives and verbal pyrotechnics abound, in pages that are anti-elegance and anti-sense, interested more in structure and echolalia than intelligent transmission. Chapter titles, such as the Kierkegaardian 'Chapter crumbs', form their own philosophical drama.

Here, the text itself is protagonist in its wordplay, tension, possibility and exuberant nonsense. In its attack on realism and rejection of simplicity, it is a modernist, difficult book, a kind of prose crust that constantly smashes together high and low, theorising on itself. Faylor is not interested in personality, story or communication, but in something else: the novelty and complexity of language and structure, the process of incorporating new elements. The everyday is made strange, and realism with its 'sickly naturalist taste' (a meta-pertinent spambot phrase) becomes the enemy.

Like the cursed videotape in *The Ring,* Faylor's kind of writing both relies on the reader's participation and changes the reader during the experience. Since these are spambot words, sense only emerges through personal association; the private meanings I derive from two phrases located near one another on the page have more to do with my own past and experiences than with whatever is in front of me, let alone in Faylor's mind.

A reliance on the reader and her ability to perceive and forge connections is required here. Meanwhile, the writer dissolves or 'ghosts'. Alienation and impersonality are preferred to the subjective 'I', which in the view of the Faylorian philosophy is a holdover from the nineteenth century. (This isn't just Faylor's idea; plenty of other writers, like Kathy Acker and Tan Lin have also explored this.) Nominally, the writing takes on certain scenes based on Faylor's life; in reality, it is totally impersonal and anti-subjective, an alternative to the inward-turning novel expressing emotional states. Are the little marks on the page signs for some greater meaning, or are they spam? Even if the latter is the case, in this world of writing, spam is never just spam. If Faylor were not Beckett, he would be Beatrice, guiding Dante through the Heaven of Spamwissenschaft. In this sense, despite its 'non-placeness', perhaps California is a theme here, or at least a certain stereotype of California. For Wittgenstein, language was the distillation of an entire form of life, and the rhythm of a book and its structure could reflect a lifestyle. Perhaps this book is an anti-product of Californian culture, a rejection of West Coast existentialism and its constant focus on the personal and emotional, which at best results in moving lyrical works, and at worst dissolves into self-pitying New Ageism. Faylor offers an abstract, fast-paced silicon alternative.

Here, unlike in the founding texts of modernism, alienation is treated not as a profound cultural malaise, but with a sort of end-of-world glee. Faylor seems to want not less but even more distance. This is a book written by someone who has read a lot of other books and doesn't want to write a parody of the past, someone for whom straightforward narrative has become dull. Faylor makes a fine gasket of his 'junk', in which machines replace humans in producing the substance of the prose, while the human self becomes a sort of editing and arranging machine.

In this post-humanist vision, the degree to which the text 'creates itself', and to which it is given shape through the organisational structures and associations of the author, becomes interesting. What is the ideal balance? And how should we even think about this? In terms of percentages (spambots do 75 per cent of the work and the human author 25 per cent)? The question of how algorithms and strategies can help us write or produce startlingly unexpected connections of words and ideas, such as 'samizdat bioflora', is an intriguing one. Faylor's work has its critical equivalent in Stanford's Franco Moretti, perhaps not coincidentally also in the area. And there are other writers doing similar work, unafraid of new developments and finding technology to be simultaneously a highly sophisticated version of hell and a source of fascination. The brain, however, may not be so happy to play along. In the morass of spambot prose, finely crafted though it may be, the eye is drawn to the subjective bits regardless, the parts written in a straightforward style that more or less makes sense. Is this something the brain does naturally, or a fault of mine in still giving so much attention to the first person 'I'? I suspect that, like me, most people are naturally drawn to writing with a lyrical tone and greater degree of comprehensibility; a series of visual postcards, with trapdoors perhaps, a highly self-conscious realism. To read Faylor is to struggle against one's own tastes, one's own 'natural' preferences. But here, once again, modernist self-interrogation has a prepared answer: Should one expect to 'enjoy' a book as one enjoys a cinnamon-sprinkled plum torte? The text is trying to do something more complex.

Just as interesting as the non-lyricism of the text, I think, is the speed of it. Sped-up prose reflects a whole philosophy, that of technology using and consuming itself. With technology, as we've all experienced, the way one reads changes: the eye glides over the spiky phrases on the page, taking in material more quickly than it would, say, Proust. While reading, the eye skips, in a self-conscious parody of the way it moves while surfing (I prefer 'scampering'?) the Web. In its gradual development and expansion from within, in its structure not pre-planned but organic and self-developing, in its 'palimpsestic' nature given to unpredictable layering, expansion and growth based on chance, *Registration Caspar* also contains an implicit element of slowness.

A rapid text created in layers takes a long time to put together in practice. One could not have made up something like this. And so, the time required to gather and arrange material forms a contrast to the frantic speed of the reading, and this is a jarring effect in itself. Faylor's sped-up prose is a kind of glitch aesthetics, in which there are too many patterns, information overload. Perhaps Faylor's own personality is what keeps the text interesting, despite his anti-subjective philosophy. What saved Beckett from the alienation of his modernism was the lyricism he injected. And what saves Faylor's text from looking like my 'Trash' folder is his conversational informality: a friendly, good-natured tone that makes just about all things possible, even the stringing together of spambot phrases.

I'm not sure that I understood all of Faylor's text, but it did make me think, especially about the speed at which one desires to live. What is preferable: a kind of ambling forward propelled by commas, linked to the speed of conversation and writing by hand? Or the kind of speed only possible after the birth of the modern computer and the techniques of cut-and-paste? Haste is intensity, compactness, a smashing together of words that accrue unexpected connotations and provoke puns owing to their collisions of ideas and language. Slowness is lyrical, visual, a smear of paint on canvas, a Gobelin tapestry work completed over years.

Just as there are people who desire to accelerate literature using technology as a motor for renovation, as in *Registration Caspar*, so there are people who hope to accelerate society as a mechanism for social destruction. What kind of people, with what kind of projects, hope to speed up society? And what kind of people, with what kind of projects, hope to slow it down? Perhaps the visual imagination is linked to a deliberate slowness, while the verbal imagination is linked to quickness, though I suspect most people have something of both. The juxtaposition of multiple speeds in a literary work — the varying rates of the process of composition, the textual confusion of the end product and the rapidity with which a reader is able to process — has an unsettling effect on the eye and brain, making one ask once again what art is, and what this work is trying to do.

Faylor's spambots would look askance at these personal speculations, but I had to know. When I asked, he confirmed my suspicion that there is something else behind the novel. 'While the Bay Area is and has been a place for the admixture of entrepreneurial ambition and artistic freedom and/or enterprise, the former has much more of a tightening grip on the area — the cost of living is soaring everywhere, as I'm sure you know, and life for artists has become ever more precarious,' he wrote to me. 'I sometimes can't help but think of *Caspar* as an angry text — and that anger definitively comes out of the economic struggles fomenting here.'

So it is. But again, this is just one element of the whole. Faylor was heavily influenced by the paintings of Mexican-born Californian artist Martín Ramírez, which are thinly stratified, segmented, mollusk-like, and suggest original ways of peeling things away, scalloped hollows and infinite readings.

*

Driving north years ago along the winding road of Highway 1, hugging the Pacific coastline, I remember thinking how easy it is in a place like this to forget that the world of technology exists at all. White foam presses up against the cliff face, and chilly beaten blue ripples in one great sheet. It's too cold to swim, it's always too cold to swim, but the bracing air smelling of salt and sandy grit feels sharp, and intense, and real. There are rock crabs snapping their little claws, birds with fantastic names (ashy storm-petrel, harlequin duck, wandering tattler, marbled murrelet) that bob about their business. The rough-and-tumble water casts black bracken and long strips of kelp onshore. One can imagine hundreds of stories of ships and mermaids emerging from this place. What seems stranger is the artificial life that also sprang up here, making this its base of operations.

One of the people who made the myth and technology of the area his concern was Jess Collins, a quiet but influential figure in Bay Area artistic history who preferred to go by 'Jess'. After working on the production of plutonium for the Manhattan Project, he grew disillusioned with his government job and turned to poetry and art, cofounding King Ubu Gallery. He would go on to create dozens of collages and paintings, including thirty-two 'translations' recomposing images from scientific works and children's books to make them more vibrant, and somehow more themselves. His work drew on George MacDonald's Scottish tales, Pythagoras's esotericism, Goethe's colour theory, Joyce's *Finnegan's Wake*, the poetry of contemporaries Robert Duncan,

Denise Levertov, Jack Spicer and James Broughton, and the events of daily life, among a host of other influences.

Visually, in many respects Jess did something similar to what Faylor does textually. Faylor's work is endogenous, operates with found material and has an unsettling 'flitting-eye effect' for the reader in the contrast between the word-heavy quickness of his prose and the palimpsestic slowness of its making. Jess's work is similarly endogenous ('the meaning appears as the painting thinks itself into being'), operates with found material, and has an unsettling effect for the viewer in the contrast between the excess visual stimuli of collage and its slow composition. But where Faylor is interested in the weird *detritus* of technology – spambots – Jess preferred to play with the weird *predecessor* of technology – Victorian myth. Faylor's works aggregate spambot 'nonsense' to draw out the ghost of the method that preceded them, while Jess' works function in reverse, compiling non-technical sources in search of a different historical outcome.

Victorian myth is not the opposite of science, but a forefather in its attempts to make sense of the world, and Jess picked up its faith that myth may have something to tell us. He wasn't against technology per se, but did think it emerged from a wider variety of sources than one might expect. In the history of visual and literary Bay Area art, surrealism, Victorian fairy tales, magic and the occult have been influences almost to the point of parody, as if to be here, in this part of the world, requires one to take a strong creative stance against the supremacy of analytical reason. Artistic life in Silicon Valley has been defined by its complex attitudes toward what the area creates. The tech world here, which arose somehow from the wild Pacific Ocean, Santa Clara fruit orchards, pale blue Santa Cruz mountains and rolling fogs of San Francisco, is not an aberration, but it does demand a thoughtful response.

Jess' final work was *Narkissos*, an enormous drawing (177.8 cm x 152.4 cm) made from graphite and gouache on cut and paste paper in a found artist's frame. It shows an enormous, beautiful lover in the foreground, eyes shut and looking down, with a black-eyed cupid with bow in hand behind him. Further back is a city building, and all around is nature, a cliff face and giant gorge with water running through, populated by mythical figures. The work has a gold-crinkled, everything-separate-yet-unified aesthetic. One can see swallows doing flips, a serpent crawling up a column, a bird's wing, a crystal, dinosaurs, a San Francisco edifice, a peacock on a branch, eyes, a comic strip, more eyes all lined up, toads, Victorian bouquets of flowers. This is a scene not of hard unsentimental nature, but of soft urbanity, and there is a gentleness to everything.

Is the city being swallowed up by nature and the myths around it? Or is it emerging, bursting forth from that nature and those myths in an endogenous development? It's not clear, and perhaps it is not meant to be. The area, constantly self-mythologizing in the form of genius 'discoveries' by figures in the industry and entrepreneurial breakthroughs, forges its own creation stories, even as it claims to be free of history and seek points of rupture from the past in the form of 'innovation'. Sensitive people have always lived here, both working for tech companies and trying to make it as artists, and the simple oppositions of the area, nature vs technology, myth vs science, are constantly being complicated and challenged.

When I visit the Bay Area, it still does not feel like my place, and I do not think it ever will. But I've come to recognize that it *is* a place — not Musil's man without qualities, but an environment fueled by its own contradictions, intensely concerned with its relationship to technology as both the source and product of the hidden underlying forces preceding it. Jess's concerns about technological and non-technological creation are still very much alive, and his work is currently on display at the SFMOMA, where Faylor works.

What Faylor and Jess are both doing, as I understand it, is

trying to grapple with the area through a deep reconsideration of the language, verbal and visual, that composes it. The piled-up quickness and excess stimuli of their work, the product of a slow process of making and contemplation, creates a strange contradiction. Sped-up prose and detailed collage, combined with a layered process of making, serve to bewilder the eye, at the same time that the brain slows down. This moment of deep processing becomes a point of origin to consider everything in an unfamiliar way, to look once more at the world and give all that one observes a new name.

Here is the poem *Just Seeing* by Robert Duncan, a Bay Area poet and the artist Jess' lifelong lover:

takes over   everywhere   before names
this taking over of sand hillock and slope
as naming takes over as seeing takes over
this green spreading upreaching thick
    fingers from their green light branching
into deep rose, into ruddy profusions

takes over from the grey ash dead colonies
    lovely the debris   the profusion the waste
here — over there too — the flowering begins
    the sea pink-before-scarlet openings
when the sun comes thru cloud cover
    there will be bees, the mass will be busy
    coming to fruit — but lovely this grey
light — the deeper grey of the old colonies
    burnd by the sun — the living thick
    members taking over   thriving

where a secret water runs
they spread out to   ripen

The Inventor

There once was an old inventor who wanted to read faster. He tried all the usual methods to increase the speed of reading, attended a private school, good university and respected postdoctorate program, paid a qualified tutor. None of it worked, and he continued to read at the same pace. But this inventor was not easily daunted. He decided that if he could not make his brain go faster, then he would accelerate his eye. With a special contact lens that pulled the pupil back and forth, his eye was made to move more quickly over the page. This was only a superficial trick, he knew, but he hoped it would be as they said about prayer: move your lips, hold your hands together and belief will follow. The book he had chosen to read, however, spent twenty pages describing the unloading of merchandise from a boat, the cases of macaroni, the tins of tomatoes, the candles in bundles, with a level of description more suitable to a clerk's log than a novel. The improved quickness of his eye was entirely unsuited to the material before him, and this or that detail spaced widely in the description, particulars that by no right should have come into contact, blurred together in his mind. In his San Francisco office, the inventor switched out his lenses for others that moved the eyes more slowly, but then he made the mistake of picking up a fast-paced thriller, and again his eye grew confused. Tired of inventing, he removed the lenses and went downstairs for his weekly class in beekeeping. There, he wound his way through the hives collecting honey at just the right speed, sticking tiny cursive labels on the jars as he made his way.

Macabre Trunk

focusing on process / Mexico

The poster for the 1936 Mexican film *The Macabre Trunk* (*El baúl macabre*) shows a man in dark glasses and fedora holding up a bloody hand in a menacing gesture, as a pulp dream of a blonde stretches out an arm to stop him. The background is a lurid combination of yellow, purple and orange, with tilted letters in graphic novel font. Splashed over the scene is a photo of a dark-haired girl, mouth covered by a white cloth, eyes dilated in terror.

It's worth lingering over the image before pressing play, for once the film has started, there's no time to reflect again with such calm. In a misguided attempt to procure the blood his dying wife needs for a transplant, a doctor kills a series of young women, going for one charming victim after the next. As he loses himself in his scientific and sentimental quest, the action takes on a momentum of its own, and the wife is almost forgotten in the bloody operations of surgical instruments. Why am I watching this? the viewer may find herself asking at this point. The initial love story and concern of the protagonist for his young wife reveal themselves to be something of a MacGuffin, since as characters they are paper-thin. Neither is there much of a detective plot, even if police do pop around to make investigations. What is communicated is something other than the relationship between people; the real hero is *process*, the carrying out of actions, the artistic performance of murder.

Blood may be sought for a nominal end goal, but in a sense this goal has ceased to matter. Indeed, process replaces result so completely that the 'do no harm' Hippocratic oath becomes a parody of itself. The physician's promise is inverted, as in his quest to provide one patient with treatment, he sacrifices a bevy of others. We discover the real heart of terror: the abstract pursuit of means that renders the end goal incomprehensible, secondary or trivial.

I can't make any claim that I'm an expert on Golden Age terror; reading the attack of the Jack-O'-Lanterns in *Goosebumps* sent me into a panic at age eight, and it took time before I started to enjoy those psychological horror stories written around the same time as the invention of the electric chair. Reading and watching the classics, however, I've come to appreciate how clever the strategies of terror are. Often, they use one of two approaches. Dispassionate process can be stripped of noble ends, so our sense of the why and wherefore of actions begins to deliquesce and rot away. Viewers become suspicious of the mechanism of plot, as with the killings in *The Macabre Trunk*.

Our scientist enjoys finding pale-faced, big-eyed, dark-haired young ladies, before carefully and lovingly doing them in, his duty converted into pleasure. Snuff films are the dark extreme of this tendency, the Oulipo of terror-as-process taken to its limit. Without even the semblance of a stopgap objective to later be forgotten, all delight is derived from pure extinguishing, crime for crime's sake.

Terror can also work by retaining faith in the final outcome, but infinitely deferring knowledge of what that outcome really is. Turn the corner, and what will you find? The particulars of the uncertain-unknown-unfathomable are niggled, and viewers enjoy the provocation. Bewitchment by the unknown isn't unique to terror, of course. Many a scientist would swear on his Bunsen burner that he entered the profession from a sense of wonder, a reaching toward territory not yet explored. In the seventeenth century, experimental scientists carried out investigations on transfusion, methods of blood circulation, and the relationship of soul to flesh.

Carried out, and were carried away. At times, perhaps the gory act of dissection was enjoyed more than purely noble pursuit warranted. The English physician Richard Lower used to visit the cells of condemned men before execution, to ask if he might have their corpses after death. He didn't want the conservative Royal Society to get them first, as it wouldn't have permitted his sort of experiments. Colleagues wrote of how unnecessarily gruesome Lower's work was, even if it did produce important results, such as his 1669 monograph *Tractatus de code item de motu et colore sanguinis* on the velocity and process of blood flow. (Ed. note: A delightful bedtime read.) Lower clearly enjoyed the process of working and writing, as his gorgeous descriptions of the vascular system make clear.

But here we are with that word again. What are we really talking about when we discuss *process*, whether in a terror film or in the circulatory system? We are speaking of technology, and our questions concerning it. Technology, or the 'science of craft', comes from the Greek τέχνη, *techne-*, 'art, skill, cunning of hand' and -λογία, *-logia*. Even if technology produces objects, it is fundamentally defined as the procedure it employs to do this. Google, Apple and other Silicon Valley tech companies may have a provisional idea of what they want to produce and certain objectives, but they're perfectly willing to swap these goals out along the way, should a more alluring proposition come along. Mark Zuckerberg may be no Mexican terror film star, but just as in *The Macabre Trunk*, the quest takes on its own momentum.

End results are shaped by process; this text logically proceeds at the speed my hand moves over the paper. If I had entered this onto a word processor directly, the results would have been different. (This, incidentally, may help explain the tone of some of those otherwise inexplicable self-help books, full of corporate jargon prizing *motivation, drive* and *entrepreneurial spirit*.) Even if an emotional engineer is a somewhat alarming idea, attitude and method are just as important as outcome, and help produce it. Technology is a means to create an end, and if the means can change along the way, so can the result. *Techne*, or craft, taps away madly on its drum, demanding attention, and we get up to dance.

Take my hand, here we go now; 1, 2, 3:

If process is what matters
let's give it attention
Boom-boom
Bad-a-bing-ba-boom-ba
Craft is all
let's give it attention
Boom-boom
Bad-a-bing-be-boom-ba
Exquisite phrase,
exquisite murder
Boom-boom
Bad-a-bing-ba-boom-bing
Lead to nothing
and that's why they matter
Boom-boom
Bad-a-bing-be-boom-ba!

Now, while we're in a good mood, it seems an appropriate moment to consider anti-utopia. How to imagine an imaginary? Easy. Think of white sands, light music, waves lapping shore. Think of emerald pools and soft fields of lilac. Think of enchanted forests, twinkling constellations, birdsong. Now think of the opposite of all this. Well? The nothing you are imagining is anti-utopia. If you are still finding this hard to imagine, perhaps it is because anti-utopia does not settle into a fixed picture-postcard image like Arcadia, but instead rushes along on the mental currents of process. One need not worry about this, for even if process is inevitable, it is also desirable, the opposite of stasis.

The dream of Arcadia and paradise is too still, unmoving as a corpse. A river of historical change that flows and flows, then suddenly debouches into a body of water surrounded by palm trees and virgins, would feel it has died, and would not be wrong.

No wonder leftist theorists worry over what will happen when the fervor of protest gives way to quietude. Odd as it may sound, the process of agitating for a new reality, if that is what one wants, may be preferable to ever actually achieving it. Elysian fields of non-activity are the opposite of revolutionary turbulence, almost bourgeois rest. Flux and possibility, the unknown and the state of change may be the true utopia.

Emphasis on process need not result in terror, necessarily. It might instead take the form of play or linguistic experimentation. An attention to process, a careful and conscientious consideration of how things are done, a shifting of noticing (I saw that flicker of your eyebrow) can be aesthetic rather than frightful. Enjoyment in process does not have to eliminate the importance of results, but it can lead one to think differently about the path to get there. And this thinking can be an active pleasure. Whether a pound of flesh or a pound of gold is found in a trunk does matter, and it would be perverse to think otherwise. But terror is process, terror is technology. And the relationship between process, participants' emotion, and end result is far more complex and recursive than may first appear.

Here, not in the random hackery of limbs, is where the true terror of such films emerges. *The Macabre Trunk* becomes the container holding Schrödinger's cat. If, as we have just said, outcome determines process, but process also affects outcome, then quantum uncertainty dictates we will not really know what is inside the trunk at the end until we open it. Perhaps the emotions of the murderer will lead him to kill more ladies than he strictly needs, filling the trunk with flesh. Perhaps the process will leave him with a feeling of distaste, and he will spare more lives than he would otherwise. Perhaps he will operate according to strict Millsian utilitarianism, and be neither generous nor sparing. Or perhaps the process of killing will change him, so he sickens of the whole business and chooses to let his beloved die, loading up the trunk with winter blankets. Utopia operates in a similar way.

[Spoiler: At the end of the film, police come to the scientist's laboratory and lift the lid. The trunk does contain decayed flesh, not directly shown but inferred by the reactions on their faces. Although we knew his supposed objective all along, the contents remain a mystery until the end.]

There was a nightmare I used to have growing up, which would vary in form. Sometimes I would begin walking up a staircase, then realize no matter how many times I lifted my right leg, my left leg, my right leg, my left leg, I was going nowhere; the staircase was infinite and I was trapped on a conduit of pure process. Other times I'd begin to walk up a staircase, right leg, left leg, right leg, left leg, enjoying the exercise, then when I reached the top would grasp I'd arrived at a different place than intended, a different place than was at the top of the stairs at the start. When I asked myself whether or not this bothered me, I did not know, and this uncertainty itself deeply disturbed me.

Again: What does this mean? Process, and the emotion that accompanies it, deserves our greatest attention. For the real terror is that long-term end results do not exist at all, and that the vague and hovering goal in the distance that gives meaning to short-term acts simply blinks out its light, showing itself to be an illusion of our addled minds. This is the fear of the religious doubter, who dreads his heaven and God are no more than figments of the imagination, will-o'-the-wisps seen in a field and mistaken for angels.

The skeptical Pascal realized this and staked his faith on the logic of process over product. The terror for Silicon Valley types is that technology will be similarly reduced to a wager of mere process, with its end goals seen to be chimeras, unimportant or false. In the *danse macabre* of fragile existence, the means we choose to fill a trunk deserve our art, philosophy and questioning just as much, if not more so, than the miracle we hope will emerge from inside. Marvellous or chilling as a situation may be, pleasure and humour can be found in events. Boom-boom, bad-a-bing-ba-boom-ba.

The End: A Snuff Film
(8 mm videotape recording, grainy)

> A taxi carried them into the foothills, past *mezcalerías* and cacti. In the plaza, they stood outside shops, looking at rugs woven by hand in different patterns and colours. None were exactly the same, though all were almost identical. Shopkeepers waited, fruitlessly for the most part, a problem of excess supply and no demand. A few tourists wandered into the tiny museum to look at exhibits on the history of stone and wool in the region. These were interesting, though not overly so, and everyone knew it was a way to kill the hours. Exposed time, waiting time. One of them felt the heat on her flesh and intimated her skin was a mere cask for the soul. She had never thought this before. But let's not get ahead of ourselves. For twenty pesos a second bought himself a small wallet, black with designs of playful dogs in colour. Across the street a white van drew up, and men in suits and sunglasses, bodyguards, jumped out. Some politician or functionary must have arrived, a third thought. In reality it was the man they would soon face. At last a tuk-tuk came and started to take them, two by two, down the desert path to a location by the dam, where goats moved in herds. As the little group increased in size with each journey, a fourth wished she had not drunk so many mezcals with chilli, salt and lime while waiting. "What is the best way to pass the time, the exposed time in our casks of skin?" the first asked herself again. We're still getting ahead of ourselves, though

not so much as before. You can do one thing, then another, then another, enjoying each action, fearing and anticipating a result, then realize your conclusions about what would happen were totally wrong. They did not yet know this, however. They lined up in the baking heat and continued to wait. A fifth picked a thistle from the brush by the dam, though he did not know if it was allowed. A sixth felt her face tanning, her hair being swept by the wind. Their backs were not against the wall, for there was no wall in that desert. It did not matter. The man was now facing them. The camera was on, recording everything. Live viewers across the world were watching. Now was the moment. The man reached in his coat pocket, felt about for the handle of the hard object. No more waiting; the time had come for it to be loaded. He smiled and pulled out the pipe. 'Oh calm down, all of you,' he said. '*Ceci n'est pas un snuff film.*'

Scientific Fictions

embracing the alien and life outside the norm / Bolivia

It's always disappointed me a little that I'm not a conspiracy theorist. I'd love to compile one of those books with glossy plastic sleeves filled with photographs and newspaper clippings, which I'd take down from the shelf not every day, because that would ruin the novelty of it, but every Tuesday, say, to flip through sightings case by case. The works of conspiracy theorists, it occurs to me, are not so different from family photo albums. Sometimes we pile up evidence not to prove any scientific theory or pseudo theory, but just to say, look, these are my people, we were here and have something to say.

For the sake of it, let the book fall open to a random page. On 19 May 2016, in the El Dorado neighborhood of Santa Cruz de la Sierra, Bolivia, a bright unidentified flying object descended in the night. From it emerged the body of a supposed alien being, seen by several witnesses. According to newspaper reports, a strong light in the sky moved downward with increasing rapidity until it crashed near a tree. Three students returning from class at that moment said they saw a strange creature with long arms and short legs, and that it dashed toward the trees the moment it saw them, disappearing into the branches.

Flip, flip, flip. Here's another good one. On 6 May 1978, at four thirty in the afternoon after a slight drizzle, the sky was grey until a sudden light moving east to west illuminated the hills. It exploded on the southern side of El Zaire hill, in the middle of the jungle between Tarija in the south of Bolivia and the Salta and Jujuy provinces of Argentina.

One picture shows newspaper reporters talking to witnesses, three peasant women in bowler hats and dresses sitting on the ground. Another shows a group of scientists linked to NASA, hiding the evidence, according to a caption. Some of the evidence is bunk, of course; some of the witnesses were confused or swayed by pre-existing beliefs. But they can't all be wrong, can they? The stack of documents points to something, or so it seems. I'll have to think about it; for now, away goes the book until next week.

*

Is a build-up of accounts like this science or pseudoscience? In some sense, it doesn't matter. The stories go far beyond the tales of credulous or drunk citizens who believe shadows passing in the sky are more than birds or airplanes, and take blurry photos that appear Photoshopped with giant red circles in tabloids the next day. Scientific proof operates according to the formulation of a hypothesis and an accumulation of supporting evidence. Pseudoscientific proof works in the same way, with pseudoevidence building to support a figment of the imagination.

I don't know if alien sightings are more common in Bolivia than other places, but quite a bit has been written about them, and the sightings are well-publicised. In an Internet forum, I came across this:

> Q: Are there UFOs in Bolivia?
>
> A: Well, we should clarify that Bolivia is a UFO hotspot, with a considerable number of UFO sightings in its eastern and western sections. The presence of physical alien bases in the Bolivian Amazon, in the Andean Range, in Lake Titicaca and the Uyuni salt desert is more than evident. But if you mean ufonauts, such as the one seen in Santa Crúz's 'El Dorado' district, we should say that these are the typical bedroom visitors, also known as 'greys'. They aren't aliens, but rather bio-robotic entities that serve aliens. They generally conduct

> examinations on men and women, regardless of their race or age, while they are asleep. Therefore, detecting them is very difficult, but they can be found in towns, cities and neighborhoods in the countryside. They are constantly monitoring humankind for dark purposes, such as human cloning.

Not a trustworthy source, clearly, but interesting, and probably closer to the spirit of the thing than an over-serious anthropological analysis would be. To complicate matters, historically false testimonies have been a conscious strategy used by the indigenous community to confuse reporters or researchers from the city with different aims, and the line between silly and intelligent is not as straightforward as may first appear. Aliens also crop up in the city, in the context of political messages, representing an alternative lifestyle. In La Paz in Bolivia, in opposition to the idea of the natural family, (joking?) posters for an 'extraterrestrial family' movement have appeared with the logo of a red ET-style finger.

Aliens aren't just creatures from another world; they also represent a stance embracing an offbeat or unconventional lifestyle outside the norm of urban life, whether this be indigenous or bohemian. Extraterrestrials may be defined by their foreign provenance, but they very much belong to the current moment.

Let's say you're looking to spot an alien. To begin with, you need to know what one looks like. This isn't necessarily easy. From watching certain television programs, I know aliens aren't always little green men, that in fact they hardly ever are. So how can we know what they *do* look like? Perhaps fiction can help create possibilities for forms they might take. A mix of local indigenous myths and urban influences alchemized into fictional material can predict what may or may not exist.

This strange world of ideas first sent its beam of light my way while translating the Bolivian writer Liliana Colanzi. Her partner, Edmundo Paz Soldán, is interested in combining pop culture with speculative worlds that transcend the regional boundaries of Latin American literature. In his stories, people are constantly dealing with technology, incorporating it into their life or dealing with its possibilities and perils in some way. Edmundo has created an alternate world called Iris, in which the human genetic structure is modified with ambiguous consequences.

Liliana, for her part, writes more realistic short stories, though her worlds are always slightly disturbed or destabilized by glimpses of horror and the fantastic. Her collection of stories, *Our Dead World*, was influenced by her doctoral work on Argentine and Brazilian science fiction, Roberto Bolaño's reflection on the fate of the avant-garde in Latin America, the bandit in João Guimarães Rosa and the animal in Philip K. Dick, but is based in the non-academic source of her personal experience.

A simultaneous fear and interest in the way the mind can be manipulated by technology permeates the work of both Liliana and Edmundo. In the 1970s, comics and novels focused on how the body was in danger from evil external forces, how people could be kidnapped or otherwise manipulated, Liliana told me. Now, a great deal of fiction focuses on how technology can be used to influence people's ideas:

> Until the 2000s, Latin American literature represented the body as the subject of operation, in which prostheses, chips and cables were inserted as a result of violence. Cyborgs were often children of the dictatorship period, a time when technology had links to torture and espionage. In this decade, the paradigm has radically shifted. Today people submit themselves voluntarily to all kinds of surgeries to transform their bodies. As Juan Terranova writes in his novel *The Flesh*, 'Our narcissism is what will turn us into cyborgs, not climate change or fewer forests.' Something similar happened with

> the figure of the hacker, so popular in the '90s: the hacker was once a rebel opposed to the neoliberal system, who used the virtual world to destabilize it from the outside. With the expansion of the cybernetic government, the neoliberal system is now so profoundly rooted in the psyche that it's no longer possible to conceive of any reality outside the system.

If technology can alter ideas, so can fiction. Contemporary writing is exploring the disturbing ways that technology can be used as a source of psychological cruelty, an exquisitely fine-tuned method for inflicting emotional damage. But this negativity is balanced by the wonderfully constructive potential use of the same tools as an imaginative source, whether as an aid to the procedure of writing or as subject matter. A playful dérive through the archive can draw on elements of colonial and nineteenth-century history, give them a fantastical spin, and suggest compelling new visions of identity and community far more complex than whatever is stamped in one's passport.

'Alien' can mean foreigner too, of course. What is the difference between an alien and a non-alien? Is it ever possible for an alien to be treated like a native, without confronting an atmosphere of suspicion? And is it possible to craft an existence for oneself that, while involving community, is lateral to how others live, so that one can simply be what one is without erroneous conclusions based on place of origin?

Alison Spedding, a British economist, anthropologist and writer living and working in Bolivia, is one of the authors whom Liliana recommended I read. Spedding was born in 1962 and studied archaeology and anthropology, and later philosophy, at King's College, Cambridge. She then received a PhD from the London School of Economics. In one of the photos I find online, Spedding wears a typical Bolivian hat, black shirt, black trousers, black leather jacket and tinted glasses. She smiles broadly, hands folded, eyes lively. In another, she looks identical but frowns theatrically. An interview notes that she thinks in a mixture of Spanish, English and indigenous Aymara, that she is intelligent and speaks incredibly quickly. Her prose reflects this, jumping imaginatively from one idea and voice to another at breakneck pace.

Let me be perfectly honest: Spedding's work is far too frenetic for me, and I'm usually drawn to narratives that are either more lyrically tender or more comic than the kind she writes. But I'm intrigued to know how Spedding came to be what she is, how she left the beaten track of her English upbringing so completely to make a new home and reality in Bolivia.

I studied at King's College myself, and still feel a deep attachment to it. Looking at the photo of Spedding and reading her work, though, I can see how she felt out of place there. King's is one of those rare locations exactly as you imagine it from pictures, exactly the same when you visit years later, exactly the stereotype others make it out to be. Towering, gorgeous chapel. Impeccable lawn. Kindly porters. Rowing competitions on the Backs. Champagne punting. Vague slight leftism. Absurd formal hall dinner menus, preceded by Latin grace. Searching through old menus at random, I came upon 'Fenland white gate farm organic squash veloute, Lincolnshire parsnip foam, organic black turtle beans' and 'Tudor Rose Fennel and candied orange cream cheese, orange blossom jelly, strawberry ice, rosewater Shrewsbury cakes'. Much of the time we were not entirely sure what we were eating.

Excellent, all of it, if you're the sort who enjoys watching *Downton Abbey* with glass of wine in hand and tongue firmly in cheek. Spedding doesn't seem that type. In an interview, she confirms that she despised Cambridge. When she was in her early twenties, she wrote a trilogy of fantasy novels set approximately at the time of Alexander the Great, which imagine an alternative history in which Alexander dies and the female protagonist, Aleizon Ailix Ayndra, goes on to fulfill Alexander's destiny. In 1989, Spedding moved overseas, and after a bit of wandering settled in Bolivia, where she began to lecture at San Andrés University in La Paz. After five years of research in the coca fields, living in the same conditions as the workers, she published her works *Wachu Wachu. Cultivation of coca and identity in the Yungas of La Paz* and *Kausachun-Coca.*

Within the country, Spedding has become known as an outspoken critic of the government's policy of cracking down on peasant coca farmers. When she was given an inordinately heavy sentence in 1998 during a police raid for possession of drugs and sent to jail, many academics considered the arrest politically motivated and campaigned for her release. Two years later, she was released on payment of a surety. Now, she's a leader in the indigenous Aymara community and grows her own coca. She has also written several books in Spanish, including the Andean picaresque novel *Manuel and Fortunato,* the thriller *The wind in the mountain range* and the sci-fi novel *Saturnina from time to time.*

Together these form a trilogy about the indigenous community from the seventeenth century to the year 2086, centered on Saturnina Mamani, an alter ego of Spedding herself. Within Bolivia, *Saturnina from time to time* has become a cult classic. Instead of an alternate past, as in the Alexander books, it creates an alternate future. The cover of the first edition shows a spaceship moving over a moon and past Saturn, against a brilliant blue sky filled with stars. Earth is nowhere to be seen. On the cover of the second edition, the spaceship flies up and over an Inca paradise in a white landscape. Living in La Paz, Spedding has likely visited the Valley of the Moon (Valle de la Luna) just outside the city, which is like no other place in the world with its barren rocky surface, dryness and lack of life. Perhaps it was even there that she dreamed up the alternate world of *Saturnina from time to time.*

Composed of thirty-four oral testimonies with women in

Qullasuyu Marka, what used to be Bolivia, the book describes how Saturnina Mamani Guarache, a legendary Andean warrior, destroyed the Martian moon of Fobos and later the Inca temple of Coricancha in Cusco. Accused of leading an anarcho-feminist organization called the Flora Tristan Command, she is arrested in Peru and accused of promoting Indian subversion in the region by the regimen of Qullasuyu. The story takes place in a bar in Ceres Orbital after the War of Liberation of 2022, where the women discuss the truth of what happened during the revolution and the cosmovision that brought them to the stars. The vision uses historical materials but is forward-looking, and despite the inclusion of archival material is representative of a myth of origin.

It is significant that this imagined future world is largely one without modern technologies. In the appendix, a 'History' of the Liberated Zone, we read:

> After the denationalization and subsequent opening of the international financial economy in 2030, and the formalization of the Syndicate in 2038, there were once again the economic and technological resources to go back to setting up broadcasts with international reach and even 3D-channels. Andean nationalism had already taken over the country to the extent that most of the population refused out of principle to have television receivers in their houses, and in not a few cases refused radios. Even in the present day, modern means of communication are considered unnecessary for the country itself and only serve to pass information to those on the outside.

Eventually these communicative technologies are restored, but they are used for other purposes, and don't seem too necessary. People find a way to communicate with each another in different and more direct ways.

There is a violence to this social vision I find deeply unattractive, yet I think Spedding's work deserves attention. Despite her particular views, what she has to say is not really political; this is an imaginative, not bureaucratic, restructuring of the world, one that plays with history but is ahistorical in its threading of perspectives. Someone with a different and slower way of being, psychologically subtle and interested in close noticing — and I think Liliana, among others, is heading in an interesting direction — could pick up these ideas to write fascinating 'false histories', ones that outstrip in interest the political polemic of the dailies a hundredfold.

Beyond genre or style, Spedding's work is evidence of a powerful vision that blurs the line between fiction and true observation, in the same way as someone telling a story about a bright light in the sky. *Saturnina from time to time* works as an imaginative founding myth for the place in which Spedding finds herself, a sophisticated allegory of the world around her that uses nationalist materials for non-nationalist ends. Through her work and her choice to make a life in Bolivia rather than England, Spedding has carved out her own world and her own place within the community, an act of bravery I admire.

Whether or not life exists on other planets, the 'alien' is a playful response to and incorporation of technology in this world. It might assume different forms, from extraterrestrial sightings to fiction that combines traditional influences with visions of futuristic communities. There's an element of language at work as well, for while it might be impossible to believe in a UFO (unidentified flying object), it might just be possible to believe in an ovni (objeto volador no identificado). This seductive power of words can persuade one to become interested in themes that otherwise would never have received a second thought, and in so doing they create new linguistic associations, new imaginative homes.

Alleged UFO sighting at King's College, Cambridge:

At midnight, I laid aside my papers and looked dreaming toward the Front Court, visible from my window in the Wilkins building. I was still an undergraduate, romantic, given over to ideas and theories rather than application. In those days, I hadn't yet lost my conviction that if an idea is beautiful enough, it must also be true.

Often I would stay up to gaze into the soft night, and it was on one such occasion that the extraordinary event happened, the one to change my life. Slowly it descended, a circle of jewel-like violet gleams surrounding a dark absence. The lights flickered slightly, on and off, a message that I should feel no fear. I wondered only if the craft had come northward over London and seen itself reflected in the city. Perhaps when looking down at our planet, other beings had viewed the lights below as a mirror of their own lights. Here now, in this college, we were far from any capital, and all was silent, at peace.

I had never given much thought to extraterrestrial visitors before, though I suppose nothing in my personal or spiritual philosophy precluded the possibility of their existence. If this specimen was representative of the rest, it seemed harmless enough. The sound it made was soft, a whir, a hum. A decorous and soothing frequency, shifting steadily between two notes. To my knowledge, such vessels have traditionally been described as round and disc-shaped, and this one fit that description too, except it also had five diamonds placed at irregular intervals around the rim. I couldn't give you its measurements in fathoms and cubits, but certainly it was smaller than our majestic chapel.

My sighting was a lucky chance. Under normal circumstances, my eye would never have turned that way at all, but I'd been puzzling over the reproduction in stone of our college coat of arms, which appears over the archway of the main entrance by the Porter's Lodge. At this hour, it was just barely visible in the lamplight. Three roses were there, *argent, barbed and seeded proper,* capped by a fleur-de-lis and lion. The shape of those flowers had always fascinated me, and I was contemplating them, on the verge of drifting off, when I saw the vessel approach.

Its violet glow preceded it, as it glided along King's Parade and turned in the direction of the College, before reaching Senate House Passage. It took the 90-degree turn without slowing, flying into the grounds low enough that I could recognize the rose shape on its underbelly. The way it moved was stately yet efficient, as if it had a task to complete.

Passing my window, the vessel came to a pause in front of the Gibbs building, where it remained hovering a moment. From the sole lighted window on the upper floor, a wizened face peered out, and appeared to make some kind of signal. It was too far to see, but I believe it involved a gesture of the hand. Then the face disappeared, and the light in the window was extinguished. The vessel continued, flying over the building and accelerating gracefully in the direction of the River Cam. Passing over the back lawn, it proceeded in the direction of the Fellows' Garden.

All this had lasted only about two minutes, but the beauty of it is seared in my imagination forever. At the time I attempted to take a photograph of the magnificent sight, but when I developed it later, all that remained was an indefinable blur. In my memory, however, the vessel exists clearly and sharply. Its form and trajectory have never left my mind. At the start what remained with me most was its underbelly, not bright itself but illuminated by the five violet-coloured diamond lights. Later, sketching the path taken by the vessel, I realized it had traced a perfect Bézier curve.

Even later it struck me that the stone fan vaulting on the

ceiling of King's College, which John Wastell completed in 1508, was also shaped to this curve. Could the vessel have appeared to that master mason as well? The ceiling has always been referred to as fan vaulting, but now I wondered if that gothic style, with its equidistantly spaced ribbing, might also represent the path taken by the spacecraft.

Could this be the vision that affected Sir Wastell so much he extended this pattern even to the Canterbury and Peterborough cathedrals? That conoid form, that rotated curved surface, were perhaps simply a twisting of the vessel's Bézier curve.

Such thoughts did not enter my head immediately, but over the past few years they've increasingly engaged me. At the time I had no inkling of the strange byways my research would eventually take. Now I am celebrated as one of the great architects of our island, with an innovative aesthetic and utilitarian contributions to modern buildings. I wonder what ideas I would have had, however, without the chance sighting of that violet-lit, rose-shaped vessel. I wonder about the gesture made by that man in the window; I wonder where else the vessel went that night. And most of all, I wonder how many others in our history have seen just what I have, but kept their silence.

Existential Puzzles

laughing at modernity and figuring out how to live with it / India

It had been so long since I last pulled out the Snakes and Ladders set that the cardboard box had warped. I'd put it away in that attic ages ago, and if the weather hadn't been the way it was, and I hadn't needed distraction, it never would've occurred to me to take it out.

When I finally did, I was relieved to see the game was just as I remembered. Everything was the way it had been, a grid of crimson over manila, ladders all blue, snakes the colours of the rainbow. When I shook the big box, a smaller one fell from it, four pieces with felt stuck to them so they'd glide better. A wooden dice with carved numbers rattled when I tossed it, before I placed the pieces at 100, 'Home' square. Straight to victory, a win for all.

If you accept the rules of the game, virtues take you one way, vices another. It's comforting: behave in Way X and you'll zip in one direction, behave in Way Y to zoom off at a different angle. A false veneer of sense, since in reality, unlike in ludo or parcheesi, the game comes down entirely to chance. The movement of the counters depends on the roll of the dice, and the snakes and ladders have no say in the matter. Who would choose to be a snake? Who would choose to be a ladder? It doesn't matter, since the identities are chosen in advance, in what's almost a caste system. Three philosophies compete in one game: the philosophy of randomness, the philosophy of virtue and vice, the philosophy of existential fixedness. No wonder the kids who play it get confused.

This situation must not be so nice for the snakes and ladders inside the game, either. Imagine a snake who generally enjoys being a snake, one who's grown up hearing that life as a ladder is dull. To have rungs is too rigid a lifestyle, Snake thinks; a squigglier style of living is preferable—more open, more free, more enjoyable in its bending of curves than the extended labor of the upward climb.

These are the ideas of Snake's elders, and he believes them. But one day, by means of what curiosity or pamphlet who knows, Snake loses interest in representing vice. He wants to be a Born-Again Ladder, wants it more than anything. Yet change is impossible; stuck in his current form, never dying or growing older, he's destined to remain what he is forever and ever. On the board, Snake is so close to what he wants, yet can never be.

Now when Snake sends the counters of the hopeful faces bent over him to perdition, perhaps even down the dreaded *Eleven Square Chute,* he feels a stinging remorse. He shimmied a small boy down that chute just this morning, and the boy cried and cried, and Snake didn't even mean to do it. This is simply the way he is made. Another thing he hates is all the grooming required; as an entity made for the 'slide', he has to keep his skin smooth and not jagged, so the counters don't get snagged during the fall.

Can he at least 'slide' down himself one day? To climb a ladder would be out of the question, since for that you need a good set of legs. Nor could a ladder ever slide down a snake, since this would crush it. Another doubt: the 'slide' is of unspecified duration. Is the climb or fall meant to last days or minutes, seconds or years? Even this is something he doesn't know. Snake suffers from excessive self-inquiry.

Next to him is his equivalent, the entity he wishes to become. Ladder represents Virtue, Honesty and Industry. She regularly sends her counters up a few notches, toward the desired golden square of Nirvana, Happiness and Worldly Success. But occasionally, just occasionally, she wishes she could do the opposite — not out of malevolence, but simply because even she can see the contradiction in what she does.

What sense is there in sending Sameer Jr, so lazy, up the ladder of Felicity, simply because he rolled a 5? Where is the karma? It's true, too, that at certain dark moments she feels herself drawn to the voluptuous downward curves of the snakes ... but how can she access that way of being? Technically it's true that a ladder can go both ways, but hers only travels in one direction: up. Is it possible for a ladder to take itself? Is she a subject, or merely a means?

These are the sorts of questions that Ladder agonizes over, confused. Faced with these conditions, Snake and Ladder, as well as their brethren, are tempted to simply fold up the board and tuck themselves back in the cupboard. Both outwardly mock the others' way of being, yet at the same time find in it a secret attraction. What to do? They peel strips of paper off the board and start to make cartoons. Snake prefers simple pen outlines, in the style of the elephant-digesting boa constrictor of *Le petit prince.* Ladder tends toward more elaborate, detailed sketches. The little bits of paper began to circulate among the other beings on the board, and on the sly both sides admire the images of the other.

Despite themselves, they laugh. The situation is evident to all, and something must happen. A convention is called. There's time for it; the ones who used to play the game have grown up almost without their noticing it, and the box has been stashed away in the attic. In the silence and darkness, the snakes and ladders don't lose their desire to talk. This desire in fact increases, as they have more time to think, and their thoughts blossom into speeches and drawings.

During the days in the attic, the First Snakes and Ladders Conference is held. It has all the importance of a Yalta or Potsdam, but the tone is more ludic. While Othello occupies himself with his out-of-alignment spine and Backgammon busies herself with the removal of undesired discs, the snakes and ladders go about conversing.

SNAKE: Let's begin by taking history into account. We are the natives of this land. Before you foreigners entered carrying your List of Rules, we were a perfectly happy country of Snakes. I admit our writhing about had little order, but with the arrival of you newcomers, a damper was placed on our national spirit. Straight lines, progress, do this and do that.

LADDER: How strange I should be playing Devil's Advocate, as what we do is bring people toward the divine. In any case, let me defend the settlers who bring progress. What would our game have amounted to without it? Not a trace of rhyme or reason, counters where they please — chaos! The form of a Ladder is also the form of railway tracks, and where would we be without good, reliable trains?

SNAKE: Developments made by a small group intended for the same — you make me hiss. Only the most absurd and aspirational among us consort with you Ladders. The rest simply laugh and circulate funny cartoons.

LADDER: Oh, we saw those. Some were quite amusing. Anyway, let's keep in mind why we are here today. Some of you Snakes are attracted to our Ladder lifestyle — and vice versa. You must begin by opening your mind. It's not just us that you Snakes don't like; you turn a skeptical eye to any Battleship or chess piece that comes to visit. All you want is to preserve your backward snake world at all costs.

SNAKE: False! We just didn't want to lose our own sense of identity before embracing other influences. Not to be rude, but please note we are more fully-formed beings than you are. We are more lithe, more colourful, more textured; we can bob our heads and our tongues can stick out straight or as pitchforks; our reptilian ancestors extend back thousands of years. We are a classical civilization ... while you, it must be said, are a bit rigid and soulless. What if we had absorbed your influences before knowing ourselves? We would have grown ill.

LADDER: Prejudices, again! What doesn't kill you, and so forth. It would have been a strong tonic, and you'd have been better off in the long run. Where is your moral fiber?

SNAKE: I must have misplaced it. We operate according to a different set of scales ... and how beautiful ours are! Our shimmering self did not die when it began to incorporate your elements, but it's not clear it improved either. It changed, lost its sense of trajectory, fractured into cubist form.

LADDER: Naturally things grew more complicated. Life is not a game of Snakes and Ladders, you know.

SNAKE: Clearly it's Monopoly. But everything would lose its charm if we thought of ourselves as silver pieces, fighting over properties and paper money ...

LADDER: Well, there we agree.

Standing with the board in hand, I notice a newspaper clipping on the back of the board, cut out with scissors and pasted on. It shows the face of a man, and there's a caption — GAGANENDRANATH TAGORE, INDIAN ARTIST. Invisible to the snakes and ladders, he presides over the debate.

Gaganendranath came from an illustrious family. He was the nephew of Rabindranath and brother of Abanindranath Tagore, and led the Bengal School of Art. In the 1920s, as the British began to bring 'modern' scientific developments into India, Gaganendranath began to work on his caricatures. According to Partha Mitter, an art historian at the University of Sussex:

> Science was not alien to India at this time. Jagadish Chandra Bose did natural sciences at Cambridge in the 1880s and began his work in the 1890s. C. V. Raman, Nobel Laureate in Physics (1930) had already been working on his research in the 1920s. Srinivasa Ramanujan, the great self-taught

mathematician died in Cambridge in 1920 and worked long before that. Gaganendranath's ambivalence is important – he belonged to a privileged westernized Bengali family. But it is important to remember that the Bengal Renaissance (which included the Tagores) represented dual heritage – modernisation of the Bengali language but also the legacy of the Enlightenment. His cartoons were a self-parody of his own class.

In watercolours painted on cardboard, which he sent as postcards to friends or published in the *Modern Revue*, Gaganendranath responded to the increasing foreign influence. When new scientific developments and modernist art entered the country, his instinctive turn was toward comedy as a useful medium to understand his confusions. He was a nationalist, and didn't like the way Western technology and science seemed to impose themselves. Especially irritating was the hypocrisy of his countrymen who wanted to uncritically embrace these developments.

In his series 'Realm of the Absurd', there is a lithograph called Moral Levitation, in which a man 'levitates' toward immoral pursuits such as drinking, smoking and entertainment, all portrayed as Western. There is also the 'metamorphosis' of an Indian man trying to hike on Western trousers, and a Bengali gentleman in dhoti who gets flak for trying to enter a train compartment holding Henri Bergson's book on laughter. In the darker 'Reform Scream' series, there are additional images showing the ways Western science works against the interests of Indians.

Before the caricature period, Gaganendranath had painted simple landscapes, inspired by 'Oriental' Japanese painting. Fields with silhouettes of palms, a faint slant of rain and scattering of birds. A man walking all alone through a landscape of blue under a moon, the colours pale and shimmering silvers and golds. The tip of a brush lowered into a glass of water and stirred it slightly, before moving across the page. Calm permeates the images, where there is almost no difference between river and shore, land and sky.

After his caricature period, when Gaganendranath began to paint again, a similar sense of calm was present, but no longer did he prefer watercolours. In a studio with his brother Abanindranath, the more 'serious' of the two, he began to look to the aestheticism of Whistler and the pre-Raphaelites, and to experiment with Western painting techniques.

His transition from negativity about Western influences, poking fun at Indian 'baboos', to picking up Western techniques like cubism, seems striking. But despite the images mocking his countrymen, he really was interested in learning from the West. It wasn't the ideas themselves that annoyed him — it was the pose, the either-or situation some of his countrymen seemed to insist upon. There was much silliness and hypocrisy in the Indian culture of the time, but Gaganendranath was gentle, not scathing; he saw clearly.

Trousers or Indian wear, French philosophy or Sri Aurobindo — the debate was a false one. Influenced by French and German developments, Gaganendranath tried his hand at a number of different styles. He painted interiors and figures alone in the landscape, city staircases and blocks of colour, women in silhouette and theatrical panels designed for magic shows.

His paintings seemed newly spiritual, using fractured angles to show the varying ways the inner being interacted with the outside world. Many objects are surrounded by an almost transparent aura, a thin layer of paint that suggests everything can be shifted just slightly left or right, up or down, to fill a different square on some imaginary grid.

A few works were directly religious, but these don't seem to be his best ones. In a cubist work called 'Resurrection', the clouds seem too fluffy, although he was not trying to make caricatures of clouds any more. (At one time, he had satirized his poet uncle Rabindranath floating across the sky toward the European cities of Paris and London, as well as the European-inspired villa of Victoria Ocampo in Argentina, where unfortunately the spiritual poet took the coquettish socialite's advances seriously.)

Gaganendranath came from an Indian Catholic background, but his Christs and candles and crosses seem too garishly obvious, props that have nothing to do with the spirit. What was the real spirit? Perhaps whatever it was that drew together the delicate wash of his watercolours, the heavy lines and full shapes of his cubist forays, the staircases and sliding panels of his city paintings, and the religious imagery of his later pictures into a complete body of work. The multiple parts form part of one game, and the spirit behind it laughs.

I can't find any record of him actually playing, but I think Gaganendranath would have liked Snakes and Ladders, especially if played on the enormous mat of some beautiful Bengali garden. The 'and' is what I like in the name of the game – snakes *and* ladders, myth *and* progress. Gaganendranath tried to defend a mythological image of his country *and* welcome new influences. Laughter as response becomes laughter as means for reconciliation and acceptance.

Let's go back now to the discussions, in which Snake is getting a bit irritated at Ladder for not understanding his critiques of technological development. He can't understand anything that prefers an A to Z route, leaving no space for luck, no room for detours.

Remember that a snake with five heads is worshipped by Hindus, and people make versions of it out of clay. You must accept snakes like this exist in order to play, even if you don't take them seriously. Chance and unexpected rolls of the dice have to be welcomed as well, even if you know your eventual goal is HOME.

LADDER: I suppose our way of seeing things *can* at times

be gray. And living with you does make things more fun. You just have to be a bit more practical, and that's where we come in. Can we agree to a truce? Any Snake or Ladder has the option to assume the other's position. Let the Spirit of Laughter animate what we do.

—On the back of the board, Gaganendranath chuckles to himself, silently–

SNAKE: That at least is compromise. We like it. Some of your developments do add something, we have to admit. Where's Dice? He was supposed to arbitrate this.

LADDER: Probably off visiting Uncle Wiggily again.

SNAKE: He'd better not invite him here for dinner. Dice can be unpredictable like that, and there's not a thing to eat ...

A hearty boom, a light titter. Standing in the half-light of the attic, board game in hand, I felt a sudden sharp pain in my ankle, then saw a shape nearly the colour of darkness slither away. Quickly, I put the board down and examined the skin: two tiny circles of blood could be seen. I thought at first of screaming, then descended the stairs calmly to the bathroom, where I cleaned the wound. Antiseptic and heavy gauze. Perhaps the Snake was just seeking a bite to eat for Dice and Uncle Wiggily? I drew a quick picture of it so I wouldn't forget, then slipped on my pyjamas and into bed. There Snakes and Ladders, Ladders and Snakes populated my mind. An immense temple of progress shot to the sky and tumbled to the ground over and over, in hundreds of ways—all the infinite variations of dream luck.

A Brief Theory of Comedy

'Life, we know too well, is not a Comedy, but something strangely mixed,' writes George Meredith in an essay. But neither, he adds, is it a 'vile mask'. Comedy can be an initial response to what is baffling, and not free of either opinion or darker sentiment.

True comedy is different from either mean-spirited satire or frivolous humor, and is simultaneously a form of critique and tenderness. 'You may estimate your capacity for Comic perception by being able to detect the ridicule of them you love, without loving them less: and more by being able to see yourself somewhat ridiculous in dear eyes, and accepting the correction their image of you proposes,' Meredith writes.

Spiritual art can emerge from the comic just as much as from melancholy. Comedy is a way of drawing the hidden elements from a situation, and operates first by dissolving tensions. Absurdity prepares one for anything, and is a way to open expectations, peel away, disclose, reveal.

Most of the time we operate with a preconceived notion of the world; comedy clears the fog and lets us see with new eyes.

Shadow Puppet

removing as a part of creation / Turkey

In the atelier, men and women bend over a table cluttered with equipment. Boxes and cables rest under fluorescent lights, and the old pop song 'Silemezler Gonlumden', by Gonul Akkor, plays on the radio. A strip of pure light falls on the Bosphorus outside, and in the distance the silhouette of a ship can be seen. Busy with work, no one looks out the window. Over the past few years, the number of those who call themselves Makers has grown steadily in Istanbul. These people want to use existing technology for personal and non-commercial uses, either out of curiosity or to address specific problems like medical conditions. Their work draws attention to the ways one can welcome a mass of creative and collaborative influences, and at the same time notice what is missing, to help create what a person or society might need.

Zeynep Karagöz, one of the Makers, wrote to me with a story close to her heart. She runs a charity called Robotel Türkiye, which aims to help children with finger or hand disabilities by building 3D printed replacements. The project took off late last year when they decided to expand through Turkey, she said. Everything began with a call from Konya, in central Anatolia. 'I've heard about what you're doing. Can you help my boy?' Immediately she said yes. At the time Anatolia was too far away, and the boy was unable to travel, so the Makers were forced to solve the problem locally. After two weeks of research with 3D printing forums, they contacted several possible volunteers to take photos and measurements; then they implemented the design and printed the hand.

The first hand made was a primitive model in different colours. When the boy, Mehmet, was given the hand to test, he came back not wearing it but holding it in a bag. The next print was a better fit. When he returned the second time still wearing it, the Makers were very happy. Mehmet told them he had tested the hand and said what he had been able to do with it, and his answers were filmed as a video. 'Did you try to ride a bike?' they asked. 'I didn't think about that. But I'll try,' he said. A week later he called to say he'd gone with his brother to Büyükada, one of the islands near Istanbul. He'd been riding all day, and had forgotten how much he missed it. 'He sounded so sincerely happy it brought tears of joy to our eyes,' said Zeynep. The feeling has stayed with them ever since, and is what makes them continue to do what they do, though it earns them no money.

Zeynep sent a video of Mehmet with his new hand. You can see him pick up and shake a bottle of hairspray, wield a hammer, steadily pour one glass of water into another. She also sent a video of a little girl named Yağmur, very slight, with long brown hair. Yağmur's hand isn't fully formed because she has ABS syndrome, found in 1 of every 1200 people. At the start of the video Yağmur draws a princess in a blue dress on a whiteboard, a character from her favourite movie *Frozen*. Then her new hand, smaller than Mehmet's and with bright red straps, is put on. Several people were needed to create it. One person invented the model, another printed and assembled it, a few more tested it to make sure it worked. Now Yağmur says that she is happy, and that she wants snowflakes on her new hand. Like Mehmet, she is able to ride a bike and play. When she gets older, a larger hand will be printed out.

*

Somewhere across the city, a curtain comes up and a pair of puppets slides into view. Behind the screen, men in caps and loose-fitting blouses move the figures with wood sticks, as they laugh quietly. A cloth has been pinned up with a rectangle cut from its center. Against the backlit paper, the dangling figures tell a story. At times the puppets move rhythmically, as if to a chant or tambourine beat. Other times they jerk abruptly, as when one of the male characters tries to imitate a female dancer's movements. Left and right, there the hips go quick. The crowd laughs, and a little girl in a pullover too big for her, with hair piled into a high ponytail of green elastic — a girl who until then had looked very serious, chin at rest on her hands — at last cracks a smile.

In the coffeehouses, gardens, palaces, and public squares of the fourteenth-century Ottoman empire, this kind of shadow play was immensely popular. The two stock figures are Karagöz and Hacivat, supposedly in opposition. Karagöz is educated and elegant, trained at a Muslim theology school and fond of sprinkling classical Turkish literary references into his speeches. Hacivat is a man of the people, rude but straightforward, with wild schemes to make money that never get off the ground. The two men have much in common, and not just in their matching black beards, puffy hats and tunics in different shades of the same style. They share a rhythm of speaking, a certain sense of humor.

Today, the shadow play is still a common sight in Istanbul. This morning a lead artist, or Haveli, will watch his apprentice carefully add touches of black paint to the hand-stitched camel and oxskin cut-outs, while he nods in approval. During the afternoon performance, this apprentice will speak his lines in a deep voice behind the screen, as shadows accompany the puppet figures. The Haveli will be pleased, yet feel something is missing. Not a body part, but something less tangible.

The number of jokes and their quickness have come to seem dizzying. One movement after another is executed, line follows on line. The words consume those that preceded them, and are chased by a dissolve of raucous cackling. Since the Haveli is a master, the puppet duo is perfectly complementary, and crafted with care. Yet, the endless banter that keeps them in movement makes him tired. What he needs is a single perfect joke to accompany his figures. Just one, to be told a single time or repeated. No more than that is necessary. The whole show that spools over a few useless hours could then take just one minute. All the jokes he hears now are simply sparks that fly off the absent Joke.

As the Haveli watches his apprentice at work, he wonders if he is any closer to the Joke than this young man. This question terrifies him, and he is paralyzed by fear. What if, seek though he may, he will never find? The Haveli is already an old man. The probability that he will die without landing his perfect joke is great. What will the little girl in the audience do if he risks telling his Joke, and it fails?

The perfect Joke becomes the constant shadow that accompanies the figures as they move and laugh across the lighted screen. The opposite of Karagöz is not Hacivat. The opposite of both of their imperfection is the perfection of the ideal joke. Behind making is absence, and to produce in an unproductive way makes one aware of what is missing in production — the *unpuppet*. There is a thing that one can never achieve, an absence one realizes is there during the process of creation.

*

As the tulip matures it sheds its petals, torquing round the central bulb. Crafted by some subtle hand, the colours are also subtle: brush strokes in orange, shading into reddish-yellow where a shadow falls. Here is a thin and jagged fringe, a slight fold, nested layers; there is a thin but firm stalk with a bit of down. Satiny petals loosen themselves from the centre.

In front of the Sultanahmet Mosque, masses of tulips stretch upward, just like the spires behind them. During Istanbul's annual tulip festival, the 'King of Bulbs' is celebrated alongside other flowers like pansies and bluebottles.

The petals of the tulip swirl inward, yet the flower links to the outside world. A gift from Holland in the sixteenth century, it soon became wildly popular, filling kitchens and bedrooms during the Tulip Renaissance, transforming into a symbol. More than a gift or aesthetic purchase, it represented something larger than society. In his essay 'The Bitter Smell of Tulips' in *Still Life with a Bridle,* Polish writer Zbigniew Herbert uses the flower to illustrate the Dutch speculative bubble:

> (...) tulipomania was a very complex phenomenon. It seems the most decisive and important aspect of the problem was economic; in other words, the order of the stock market was introduced into the order of nature. The tulip began to lose the properties and charms of a flower: it grew pale, lost its colours and shapes, became an abstraction, a name, a symbol interchangeable with a certain amount of money. Complicated tables existed on which individual varieties were arranged according to the changing market prices like valuable papers or money rates.

Someone who walks through this garden, fond of noticing shapes, will see the geometry of the flower right away. Self-conscious about appreciating flowers in an age of cellophane, it will take her longer than it once did before she at last succumbs to the flower's perfume. She observes the form of the flower, giving it time to reveal its subtleties as it rests in its bed before the mosque. The hollowed-out shape of the tulip is the shape of a bubble, she thinks. Beautiful petals that enclose nothing. Finally giving in, she leans over and breathes in. But this perfect flower has no scent at all. Do the gardeners, the makers of tulips, prefer such modesty?

Tending to flowers was in other times the responsibility of guildsmen. In many ways the Maker culture is an updated version of the guild of medieval times, in which knowledge is shared but each artisan has an area of expertise. The medieval Turkish guild (*esnaf*) was a form of organisation that got rid, or took care, of 'basic' needs, a name for something not basic at all. Basic needs are an array of functions and chaos of random necessities that may or may not relate to one's desires. In the *esnaf*, basic needs such as social security, low interest credit, price monitoring and work distribution were taken care of, so that one could focus on one's own craft, whatever this might be. Absence turns to presence.

*

The lathe is a traditional tool that works by getting rid of excess content. It turns and transforms, spins and spins. A workpiece is cut and sanded, knurled and dulled, turned and burnished. The material gets smaller as it is whittled, all excess sheared away.

New inventions can similarly be created in an informal environment of shared knowledge. Given a chaos of inputs and influences, the lathe of the mind can cull meaning from a mass of material that seems overwhelming. In this sense, perhaps the Maker movement can be called the unmaker movement. A good part of making involves unmaking, editing, stripping down. Experience before comprehension. Doing a lot then culling and rearranging until what remains is the heart. How can one best understand the relationship between excess and essence, twirling it into shape?

In a field of tulips, each might be perfectly crafted, yet form part of a larger field able to communicate with itself. Mehmet and Yağmur spin through the garden on their bicycles; laughter is heard from a coffeehouse in another century.

Bird towel procession

In an old painting, the makers of bath towels walk by in a

guild parade. Each one carries a towel, and on top of each towel is a bird. The towels look like flattened versions of the birds that sit above them, and are in the same colours. These colours are mirrored in the clothing of the crowd that watches and says nothing, yet wonders what the towels are for. Bird: towels. Bird: other birds (variations). Bird: itself. Muscles emerge from torsos. Wings fold back. Feathers hang in a particular way. I feel the bird towel in my arms, but even closer the presence of the bird not here, the perfect and absent bird toward which all of us move, a fluttering in our hearts.

Magic Lantern

transforming sense data to produce effects / France

The bird in its golden cage stretches out its thin legs and puffs out its chest. *Piou,* it says. *Piou-piou.* This song was taught to it by a lady with a real canary, and the sequence was repeated over and over until learned. *Piou. Piou-piou.* Standing discreetly to one side of the room in his crimson dressing gown, the magician takes a few grand steps forward, sweeping his arm in an arc, as the bird in its golden cage repeats the sequence a final time. The spectators know it is not a real animal, but this does not bother them; in fact, it dazzles them.

The bird is a descendant of The Flying Pigeon, a mechanical bird developed in ancient Greece, and the first robot of all time, for those of an anachronistic bent. It is one of a long line of artificial birds, used throughout history to better understand the natural biological body of animals. During the Renaissance, even da Vinci dedicated himself to creating mechanical ducks. In the nineteenth century, the bird was picked up by magicians. Of these performers, none succeeded in so many spectacular variations as Robert-Houdin. His mechanical birds were better learners than those of any other, and much beloved by his audiences. Best of all, since the birds were never faulty, his shows could always go off without a hitch.

Now Robert-Houdin draws the bird from its cage, cradling it with one glove-covered palm and cupping with the other as if it will fly away. With a single rapid gesture, he removes his hand to display it, and it's gone! The public applauds, loudly, for a long time.

Robert-Houdin, the magician, appears as a main figure in *Topografía de lo insólito,* the Spanish version of an essay by writer Chloe Aridjis, published by Fondo de Cultura Económica. In

Chloe's book, the techniques of late eighteenth-century Belgian magician Etienne-Gaspard Robertson and nineteenth-century French magician Jean-Eugène Robert-Houdin are discussed alongside those of poets Gérard de Nerval, Honoré de Balzac and Arthur Rimbaud.

For Chloe, thinking is related to seeing, and there is a visual intelligence and knowledge to be acquired from sense perceptions. These perceptions pass through the mind, and are transformed by the imagination into magic, whether literally in the form of a magic show, or as an idea, historical narrative or poem. The magician, historian or poet plays the middleman role of the imagination for others, working sense data into spectacle, narrative or crafted literary form. In the meantime, the audience finds pleasure in the knowledge a gap exists between what is 'real' and what is wondrously experienced.

I copied out this list of phrases while reading:

> Poem as concrete model for mental process. Instability and defamiliarization. Back-and-forth, faith and disbelief. Doubt created by moments of terror. Aristocratic spectacles of the macabre attended by ladies. *Le bouleversement de tous les sens*. Magic lanterns and optical illusions. Mysterium, tremendum, fascinosum. Goya's Caprichos. False expectations, awaiting one thing and experiencing another. Athanasius Kircher and his hierogylphs. Rupture, travel, development. Mechanical trapezes. Miniature pastry chefs. Seances and hermetic philosophy. Fluidity and fragmentation, elixirs and chimeras. Objects, colours, contours, ideas. Papillotage. Charlatans and self-proclaimed messiahs. Disappearances and returns. Adieu, hello.

Chloe was born in New York City, but has lived in a number of places. Her mother was a translator, her father a poet-diplomat. (Later, she would translate his childhood memoir *The Child*

Poet.) She spent her early years in the Netherlands and moved to Mexico City when she was eight, where as a teenager she frequented the local goth clubs and spent her time reading widely, in both Spanish and English. From there she went to Harvard and then Oxford, followed by six years in Berlin before making her home in London.

All this moving around has given Chloe a certain way of seeing. Her books are about everyday noticing, and in the observer role adopted by inclination or circumstance, she describes phenomena that might not call the attention of others. Art appears frequently in her pages since, like a foreign city, a painting can provide material for careful looking and noticing. Within the density of visual information, Chloe discovers small clues:

Upon seeing me at the painting, Daniel came over with his notebook under his arm and asked whether I had spotted the comet, to which I said no, startled by both the thought of a comet in a painting and the fact I had missed it. I leaned closer in to scour the sky — gradation of light pink and blue thinning into yellow, like a molten version of rock sediment, dolomite, limestone, sandstone and shale, and finally found the comet. A simple white brushstroke: one milky line at the top, hardly visible.

Drawing on both her life and her imagination, Chloe is influenced by Walser, Kafka, Bernhard, Gogol. In the still lifes of her highly visual novels, almost nothing happens, but cracks start to spider over the smooth surface, and quiet tedium gives way to violence, or threatens to. *Book of Clouds* is about a young woman's life in Berlin, and her other book *Asunder* is about a young woman who works at London's National Gallery. Chloe is now finishing a novel about the Mexico of her adolescence, which moves between the Distrito Federal and Oaxaca, and which still doesn't have a name.

When we chatted, Chloe had just come back from The British

Library, a magical place in its own right. (The last time I entered its hushed spaces, I listened to a man writing a book about his ancestors request multiple volumes on falconry, and wandered through an exhibition on the beat poets.) We talked about the relationship between her works and the Work, and about the strangeness of the imagination. In the background Chloe's 'very demanding' cat, Ludwig, purred away. He was rescued three summers ago from a Greek island, and she still can't decide whether he's a Beethoven or a Wittgenstein or the mad King Ludwig of Bavaria. Here's what she said, with my questions removed:

> I have no interest in memoir, but I like to shape material from the past or present into new kinds of narrative, defamiliarizing it. The idea of defamiliarization interests me, seeing everything as slightly askew, in a reality that's a little off-kilter. A foreigner has that perspective already; you see things from the outside. I imagine things in a parallel universe: how they would exist there, what a person would represent, what their inner world might be.
>
> Like Adam Smith's sympathetic observer seeing yourself from above, exactly. An idea of ideas that comes only from a certain level of abstraction or distancing. In Berlin it was easy because I was living in a foreign language, and though I spoke German, I didn't feel attached to it. It allowed me to go much deeper into my thoughts, understand scenarios, study faces on public transportation, imagine what other people's stories might be.
>
> You can do that in any city, of course; even the countryside. But there was something very particular about Berlin. It had a strangeness to it. A lot of the London of the past, 80s Camden Town and the goths, has that, too; though that side of London isn't there much any more. There are aging goths and people from that time, but it's not the same. London is more of a

corporate city. It has eccentric characters too, strange and extravagant ones, but not so many as in Berlin. There's a different set of tensions, less interesting to me. There's a charm in that decay, but also always a risk of disenchantment.

In every place there are things to find; you just have to be more open to them here. In Berlin they're handed to you on a platter, though the city has changed a lot in the last decade. I'm extremely happy and grateful I was living in Berlin when I wrote my first novel, that that's where it was set. I just went around with my notebook. With this new novel, in a way it was even more challenging, because I don't live in Mexico and haven't for a long time, though I spend two months of the year there. In ways it almost feels like historical fiction since it's set in the late 80s, so even the Mexico I'm writing about doesn't exist now.

The first half is set in Mexico City, the second half takes place in Oaxaca on the Pacific Coast. I invented a few new characters, and gave some of the nocturnal creatures from when I used to go out more importance than they had in my life, pulled them into focus. The one episode that really happened was when I was sixteen and ran away with a boy to Playa Zipolite. My father came looking for me and, well, it was a disaster. So there's a build-up to the 'Fugue' section, then I run away with the boy to the beach. Then there's a more fragmented, dreamlike section once I get there, and I'm completely adrift ...

Visiting to research the book was tricky. I only went back to places or avenues that are almost the same — the Zócalo, the Centro, the Catedral — to soak up the general atmosphere. You have to be careful not to write over the memories, as sometimes it can inhibit the imagination. Sometimes the places I write about are completely different now and don't exist any more, and seeing a different venue or shop does a disservice to the memory and doesn't help the story.

The funny thing is that sometimes friends think they're characters when they're not, and think they inspired a character when they haven't. It's almost the reverse of people complaining: Why did you put me in your novel? 'No,' they ask. 'Did I inspire that?' In *Asunder* there was no particular museum guard I based myself on. I interviewed several and spent a lot of time wandering through the National Gallery, and initially I thought one would give me an idea for a character. In the end, it was an amalgam of many. The narrator's best friend, Daniel, is a poet, and his identity isn't bound up with being a guard. The young woman is also completely fictional.

Including the suffragettes seemed an obvious choice, because the character shows traces of revolt against her own passivity and the passivity of her life, and the suffragettes were very passionate and anything but passive, fighting their roles. It worked well as an opposition to the girl's condition.

Often when I'm writing, in the early stages I have a constellation of themes I'd like to have in the book, and then it's a question of figuring out how they work together and resonate. With the suffragettes, the craquelure, the museum, the poets, the chatelain toward the end of the book, it all fell into place and I was very happy. But there's always a moment of great uncertainty when you're trying to map it out.

My favorite French poets have always been Baudelaire, Nerval and Mallarmé. When I got to Oxford, originally I was going to do German and write on Thomas Mann, but it was too daunting. So I switched from German to French and landed with the absolutely wonderful professor Malcolm Bowie, and began to focus on magic and the literary *fantastique* in nineteenth-century France.

There used to exist in Paris, now unfortunately it's closed, a little bookstore that my father would always visit, called the Librairie du Spectacle. It had books on clowns, pantomimes, circuses, magic shows, all kinds of performing arts. I came across references to Robert-Houdin in my early research, when my father got me Robert-Houdin's autobiography at the bookstore, and then I came across the autobiography of Étienne-Gaspard Robert, the Belgian magician with his magic lantern. And I was very struck when I began to read their autobiographies how these were bildungsromans, very much coming-of-age books, but at the same time they spoke about their craft in a way very similar to the way some poets spoke about theirs.

There began to be so many similarities with the writers I was reading, especially Mallarmé, but also Rimbaud and Baudelaire. They saw poetry as a sort of alchemical or performative act, the empty page as a stage, the verb as transformative, the city as a place to be coded and decoded around them. There was a phantasmagorical quality to life amidst these characters. It was a wonderful way to spend my twenties. I knew I probably didn't want to go into academia, but I had an amazing professor and all of this to think about.

I love poetry, but maybe poets are by personality more performative, whereas someone quieter will produce a different kind of work. A novel is a much longer endeavor. Even engaging with it is a longer act. Magic tricks are like working models of mental processes, and there's a kind of distillation going on. In mysterious ways, though, all that material from my dissertation is still in the background. That's the wonderful thing about how brains work; you never know how something from ten or fifteen years ago will enter your writing. A thought or an image can make its way into work a decade later.

The question of contemplation and distraction is interesting. In some ways I think it's one of the main issues today. I just got my first iPhone last year, and resisted. Even now, I mostly switch it off when I go out, to feel more free. I think of the days in Berlin going around with my notebook. I had one of those old-fashioned clam phones you could just snap open and shut. I can't imagine what it would have been like if I'd been constantly checking things online.

There's a certain ambivalence toward technology at the end of *Topografía*. A mention of metallic shivers. Even at the library, there are days I feel focused, but am surrounded by people texting or on Facebook. They can spend their hours however they want, but somehow it ruins the atmosphere, and you feel a pact has been broken. One goes to the library, at least I do, to work better, and it's very distracting if there's someone reaching for their phone every two minutes. There are the hand movements, the eyes darting back and forth ...

It's very important to have communities and exchanges of ideas, though. Mallarmé had Tuesday salons at his house, and serialized literature also had that element of participation with an audience. Collaboration in itself isn't a bad thing. It's just the ability to have those quiet moments on one's own, those mental silences, that I see people struggling more to have.

Recently I've collaborated on two projects, and there's a wonderful sense of being lifted outside yourself, or not being so completely channeled and funneled into your own thoughts. Last year when I co-curated the exhibition at Tate Liverpool, there was a year of emails and working with other curators, being aware of their thought processes and how they came up with narratives of the exhibition. That was a very positive experience. It was the first time in my life I'd really collaborated, and it wasn't with my writing.

Leonora Carrington was a family friend, and we would go to her house every Sunday for tea. I got to know her and her work very well. I wrote one or two shorter texts about her, then Tate Liverpool decided to do an exhibition and asked if I wanted to get involved, which was wonderful. Most of my friendships here are in the art world. I do have close writer friends, but socialize much more with the artists, and go to openings a lot. Even though I don't like much contemporary art, I still go for anthropological reasons or morbid curiosity. It's a carnival.

I think that, consciously or not, if one has the luxury of choosing one's profession, in the end people choose what is best suited to their temperament. I'm quite solitary, so I can't ever imagine being surrounded by people all day. In the afternoons I like to leave the house and go to the library for a few hours, on my own terms, and I don't speak very much to anyone.

Nobody really knows where I'm from. In a way it's liberating, but I also sort of fall through the cracks. I'm not considered Mexican in Mexico because I write in English and haven't lived there for so long, and I'm not at all on the radar in the United States. England does feel like home and I'm part of a community, I just got my British citizenship, but culturally I'm not British. I really feel like I belong to a country that doesn't exist.

A shadow with the shape of a kite moves over the ground; a light flickers. The human imagination takes sense data and transforms it into magic. But this same data can also turn to horror. Phantom presences, fears of disintegration: the imagination is able to compress all the diffuse things seen into a single hard kernel of meaning. At the heart of both magic and horror is the conviction that everything is an allusion of importance or metaphor, that no movement is inconsequential.

With the same suddenness, ways of coming together can also become ways of coming apart. Take painting, for instance. To destroy a painting, if you are so inclined, you can rush at the work with knife in hand, as suffragette Mary Richardson did with the *Rokeby Venus*. (The museum guard in *Asunder* is fascinated by this act.)

Alternatively, you can simply let time take its toll. Think of the black cracks spidering over the ivory faces of a Botticelli or Tiepolo in Rome, destruction developing in the same way the appearance of an ideal milky line might: partly as a result of chance, not entirely. Paintings crack up depending on oils, damp, location on the wall and other concerns restoration specialists make it their business to worry about. (Chloe's museum guard also takes a deep interest in craquelure.) If a brush is dragged at a 45-degree angle across a canvas, perhaps the canvas will be more likely to erode that way. Creation and destruction, the making of the milky white line and eventual cracking of the canvas may be linked.

Robert-Houdin and his mechanical bird have left the stage, and now the oval lady enters. She is made of tiny parts specially locked together, sheets of gleaming silver metal, miniature cogwheels, sapphires for the eyes and twisted bobby pins for legs. There is a golden key in her back, so you can wind her up and make her turn. This is a lady out of dreams, the most beautiful ever seen, and just to keep things interesting she has the head of a horse.

I am the magician onstage now, and to open the box I press a gold button. The oval springs up and the lady starts to travel, back and forth, back and forth, along a curve. There she sings, in her *boîte à cheval danser*, a replica of those from famous French and Swiss houses. Heel up and heel down the little thing dances, and I can make her start and stop. She doesn't have any free will by the looks of it, and her lacquered shoes go on tapping several minutes. Then the lid claps shut like an oyster shell. *C'est tout.*

Leaving her upbringing in England, the artist Leonora Carrington came to Mexico City in 1942 after a chaotic life with Max Ernst, nervous collapse and marriage to Mexican diplomat Renato Leduc. This is where she would settle, continuing her work as a writer and artist, and getting to know many of those now considered surrealists, sharing domestic spaces, jokes and experiences of motherhood.

In Mexico she wrote several story collections, including *The Oval Lady*, as well as her novel *The Hearing Trumpet*, about a 92-year-old woman who overhears her family plotting to commit her to an asylum. This is not any institution, but a place with buildings shaped like igloos and birthday cakes, with lots of friendly people willing to hear out her silly ideas. Curious and open-minded, with a sense of humor, she can get away with being a little bit crazy, connecting everyday things in odd ways, and discovering the weird links and hidden situational puns fusing different tectonic plates of experience.

In addition to curating the exhibition of Leonora's paintings, Chloe has sprinkled oblique references to the hearing trumpet throughout her work. The young woman in *Book of Clouds* wanders through the market and comes upon a sound machine ('plastic contraption shaped like a seashell'), and the old historian the woman works for as amanuensis unveils a mechanical ear.

Like Leonora in her work, Chloe touches on themes of art, private worlds and life abroad that interest her and intersect with her own life. But her world is a darker one than Leonora's, and more rational; though it yearns for the unthinking magic and humor of the surrealists. It hints that a benevolent existence devoted to painting and friendship can be both highly attractive, and a surface with cracks running over it. What hides under those surfaces to enable a world of such fantasy? Even Leonora's stories are always tales of horror.

Maybe this has to do with personalities, maybe situations. To live abroad is always to create a crack in one's settled world, and the act of self-consciously moving is already an artifice. To travel is to find a new domestic space abroad, and if one is female, to seek those spaces as neither mother nor femme fatale. But I suspect the type of personality that likes to read novels and look at paintings for a long time is also the kind drawn to that sort of life, with all its complications and delights.

Technology troubles and transforms the surfaces of reality, too. Perhaps this is unavoidable, and perhaps it is not necessarily a bad thing, as it can enrich the connections between objects and our relationship to them. But the cracks and ruptures in this surface do occasionally make one long for a more natural existence. The young woman in *Book of Clouds* joins an organization dedicated to fighting the spread of artificial light, which it sees as pollution. She writes:

> For a long time I had deplored the human fixation on light, or rather, on artificial light, even before learning the German word *Entzauberung* and agreeing with all those poets and philosophers who warned about modernity and technology intruding further and further upon the imagination. Well, I was now witness to a very serious sort of disenchantment, the disenchantment of night, when each day at around six or seven, depending on the season, dusk fell and the mania for lighting up the sky and denying darkness began. Night would never be mystical again, at least not in the city, and I sometimes had fantasies of flying through town and smashing every lit bulb, or at least those screwed into the impertinent lamps on my street, obliterating those bright and irksome reminders of the rest of humanity, if only for a few hours, before morning rose and everything revved back to life.

Chloe's narrators, intelligent and curious, if also shy and occasionally irritable, are always looking to cut down on excess stimuli, in search of an ideal state of perfect stillness and concentration. In her work, the human imagination works its best within the parameters of either a specific situation (i.e. small community) or reduced set of influences (i.e. quiet place). Such conditions are needed for intensity rather than diffusion to prevail.

Here, despite the yearning for nature, artifice can help. Whether this means a specific stage set-up for a magician, formal linguistic constraints for a poet, or a focus on a limited period of time for a historian, there are ways to imitate the reduction of stimuli. Technology can help create the conditions for this artifice, applied at the stage of either perception of sense data or imaginative process.

Perhaps one can think of the mind as something like the magic lantern so dear to the magicians Chloe describes, projecting the light of the imagination through pictures painted on glass. Technology can play with the images on the glass, or change the strength of the light. Either way, it does so with the aim that, when the image projected on the screen in the distance does appear, it is large, and clear, and illuminated.

The Little Earthquake

> I am sitting, cross-legged, on a very small square of sidewalk, almost no larger than I am. It has a crack in it, a hair-thin line. No, not just a crack: I see now that there are ruptures everywhere in the ground, very fine. The crust cracks in solid plates here, peels away in gray flakes there. Fissures of varying width run over the sidewalk, as if carved out by a toothpick or cleaver, and green grass pokes up between them, having somehow grown and survived under the surface. The edge of the rubble is made of dry bits and knobs, some smooth and others textured, and none of the cracks are exactly the same. How did they get here? Was there an earthquake? Are we sitting on a fault line? No, there was never a quack of such magnitude in this place. Quake, I mean! Ah, a typo. But you know, charlatans of earthquakes could really exist, couldn't they? People who sit on very small squares of sidewalk and convince others that the smallness itself makes a tension where they are. All these cracks and splinters show it will

soon burst. Yes, any minute! This square of sidewalk will go flying to pieces, and the beast from beneath will rise. Trembling, I wedge my pencil eraser under another bit of cracked earth, then lever it up until it goes flying. A black gash opens and the crack expands. I push the eraser in a little deeper, and the darkness expands even more. Just one more time ... The next person to walk by this place will see the rupture. Perhaps he will think a sledge or jackhammer was responsible. Construction work is always going on in this city. By then, I will long since have vanished, underground with the creature, or fled elsewhere.

Warp of a Field

folding conceptual topos of history / New Zealand

A star sends its light through space, and passes through the strong gravitational field of the sun. The field bends the light, so the position of the star changes. During a total solar eclipse, the difference is visible if you're standing in the right place, as Arthur Eddington was in May 1919. He'd traveled to the Island of Principe to record the disparity and prove Albert Einstein's theory of relativity was more than the fantasy of an imaginative German. In *Science and the Unseen World,* which he wrote that same year, Eddington took the long view, describing how 'centres of condensation' can build up over time:

> The years rolled by, million after million. Slight aggregations occurring casually in one place and another drew to themselves more and more particles. They warred for sovereignty, won and lost their spoil, until the matter was collected round centres of condensation, leaving vast empty spaces from which it had ebbed away. Thus gravitation slowly parted the primeval chaos. These first divisions were not the stars, but what we should call 'island universes', each ultimately to be a system of some thousands of millions of stars.

Can moments in human time operate like these centres of condensation? Are changes in emphasis capable of altering the conceptual plane? Can a group of eccentrics with certain ideas warp human history, in the same way that stars and particles of matter can warp space?

'... Take a deep breath,' Richard Pearse tells himself. He's about to fly over New Zealand, the first in the world to operate a heavier-than-air machine. He's only invented it recently, and to this point has used it exclusively for short-distance maneuvers. Today things will change. There he is now, up in the air, moving along the flight path. He notes differences in the terrain. Heading southwest, Wellington to Christchurch, he follows the great spine of the Southern Alps, hits Ashburton and continues on. To the west is Queenstown, with The Remarkables on Devil's Staircase. Below is his intended destination, Oamaru on the South Island.

His airplane crosses overlapping tectonic plates, moving from regions of dryness to those of rain. With great care, he follows the plan mapped out before leaving, not risking a single deviation or veering off course one iota. This is necessary, for were he to do so, he'd be flying blind. With the plan, he fears nothing. Maneuvering his plane through the sky, he follows the set path.

Gradual shifts are evident. The trees start to become a bit more or less yellow, mountains clump together with increasing or decreasing frequency, lakes and clouds rise or fall in number. Nor are the changes simply a matter of quantity. There are also changes in (a) quality, such as intensity of blueness; (b) relations with neighboring objects, such as hugging or torquing away; (c) attitude, such as rest or attack position; and (d) curvature, such as the way water bends around islands, so it looks like an embrace or gawping amoeboid mouth. In general, these changes in category adhere quite closely to the Aristotelian praedicamenta, and the thought occurs to Pearse that if Aristotle were an aviator today, he'd be quite the flying ace.

Peering down at Earth in this way is fascinating, but Pearse also realizes that the way he sees depends on his position in the aerial survey. Kilometer after kilometer he moves along, but were he to shift either horizontally or vertically, the attributes noted would be distinct. So he faithfully continues to follow his flight path, better not to risk a crash.

After landing his plane with care in an open field, Richard Pearse steps down, brushes the dust from his high-waisted trousers and makes his way to the center of Oamaru, a little town with colonial buildings in the neoclassical style, fronted by a gorgeous blue harbor.

An innocent bystander glances at him as he approaches. From the way he stares, Richard Pearse suspects he must look like a monster in his pilot's cap, fitted jacket, dark smoky goggles. But he's not a monster, just an aviator, one not yet known. His brown hair curls more than usual in the humidity and his cheeks shine. He looks like he needs some nice cool refreshment, a Foxton Fizz.

That's what the innocent bystander thinks, as his feet carry him ever closer to the man. One wouldn't be amiss for him either, to cool his feverish brain. The innocent bystander knows a man dressed like this in modern New Zealand isn't necessarily a real aviator, but more likely an eccentric and harmless aficionado of the growing steampunk presence the town has seen in recent years.

Modernized Victorian dress complete with suspenders, Inverness jackets, leather spats and aviator goggles, is longer an outlandish sight in the streets. The innocent bystander slaps the man on the shoulder. 'Do you know, for a minute I thought you were Richard Pearse!' he says, laughing. 'I *am* Richard Pearse,' squeaks the man, disturbed, but the innocent bystander has already gone on, toward the town's central square.

When he gets there he sees a group of people gathered, all dressed like the man he's just passed. Ah, so I was right then, the innocent bystander thinks. What strikes him first about these people is their attitude, defiant, sly, but also amused, as if at any moment something might happen and they'll go with it, without feeling flustered. Oamaru, they say, is the steampunk capital of the world.

Notebook in hand, the innocent bystander approaches a woman with a monocle, flaring dress and umbrella, and asks about the spectacle. 'Is it ever possible to explain a mystery without destroying its allure?' the woman shoots back, half coquetting, half irritated. Then she shrugs. 'We like heaviness.' 'But why?' the innocent bystander asks, insistent. 'Isn't that a negative quality?' He thinks of lethargy, dreamless sleep, an anchor on the soul, rigid system, shackles clamped tight.

The woman looks at the innocent bystander, hands on her hips. 'Well, sir, to most of the world, it probably is. But don't you think lightness has its downside, too? It's true most technologies in our modern world promise it, and most perceive this as a liberation—'

'Yes, a liberation! That's the very word,' the innocent bystander cuts in. 'An intangibility of data, an invisibility that seems to approach some definition of the spiritual ...'

The woman holds up her hand, looks at the innocent bystander coldly. 'Don't interrupt,' she says, lowering her palm and using it to smooth the lace frills of her skirt. 'That very lightness of which you speak so highly can easily begin to seem oppressive. All of us came here in part because of this. Our bones started to feel brittle and light, and a creeping fear that we'd drift away, bodiless, afflicted us. We felt we were becoming air dissolved in air, so we took steps... but sit down and let me explain.'

Overwhelmed, the innocent bystander finds a place on a bench to listen, as the woman takes a long, cool sip of her Foxton Fizz, and begins to speak.

*

The future, they say, is the upload of the self onto a cloud, but do we really desire a life so mental? Esprit, lightness, was the basis for Talleyrand's ideas of dispensability, in which humans, bits of debris, float helpless down the stream of history. Heaviness, in contrast, is density, solidity, continuity, reliability, loyalty and tradition, all of which feel so very much more real. How conservative, one might moan; but no, not at all.

This heaviness requires a certain courage, the assumption of a stand, just as a stick planted in the mud of a stream must hold strong as water rushes round it. In the same way a stick can alter the flow of the water, so a revaluation of heaviness can begin to warp values. Think of the *fin-de-siècle* appreciation of stasis, with its description of dead flowers, Japanese vases and corpses, occasionally even arriving at the level of necrophilia. We don't want to go that far, but we do see value in the solidity of objects.

The woman sticks her hand in a fold of her skirts and comes up with something metallic: not a brooch, but a pair of goggles.

History is an exquisitely thin cloth, receptive to the slightest touch. Any symbol dropped on it will carry weight, just as a heavy brooch pulls down the conceptual fabric. The fabric of time is itself a fabrication, and those of us here, deliberately or not, are creating an alternate history, pulling the cloth our way. What is required is the introduction of a subtle change, moving past into present, or present into past, to create a new conceptual flight path.

This might seem very abstract, but in practice it is not at all. Changing ideas comes down to objects, dressing-up, works of art, writing. Sometimes all this play remains just play. But sometimes new topologies of knowledge are created, vivid enough to affect the 'real'. The brooch falls on the cloth, the cloth falls on the ground in folds, a warp is created. The flight path changes. Even if the narrative of progress still pilots the craft, the path is no longer linear.

With a cool gesture, the woman slips on the goggles.

Imagine the new air routes that will exist. Will the planes crash? Will they collide? Will they go up in flames? We do not know. What we do know is that when history is tinkered with in this way, strange things can occur. The un-happened can happen, the flat fabric of history can warp, the strange can transpire. Glitches can result from the interior of the new folds, the brushing together of cloth from disparate parts of the topography.

Given this, two attitudes appear as options. Lightness can be taken seriously, in a sacred and deliberate abandonment of the past. Or, heaviness can be worn with high spirits, drawing from the archive of history in a whimsical way. Those of us here have chosen the second option. We love history and are not prepared to dispose of it. That would simply be too great a loss to bear.

As I said, a heavy mantle can provide greater freedom than an unmoored weightlessness, for without it, we are reduced to what we do right now. Those of us here believe in ideas and interiors, we believe in history, we believe in the mind, we believe in those dear to us, we believe in stories.

The woman pours out the rest of her Foxton Fizz and walks away, holding the glass bottle.

*

Brief aside: Far away, at the Université Paris Diderot, the Italian mathematician Olivia Caramello updates her website. At the start of her career, she writes, she 'had the intuition Grothendieck topoi could serve as sorts of 'bridges' for effectively transferring information between distinct mathematical theories'. Her work on Topos- Theoretic Bridges serves as an attempt toward a unified theory of the universe, and it is controversial; although not so much for the content itself, as for its authorship: her previous professors stake a claim to the work.

Since I, the one writing this, don't understand the mathematics behind the theory at a deep level, I imagine the bridges she describes perhaps too literally, in colours. Midnight blue, dusty rose, marigold yellow, forest green. The bridges are abstract, I know, but even the abstract can take on a visual quality. I wonder if bridges this beautiful can really exist, or if they're simply another kind of poetry, one with a particular appeal for the logical mind.

Still flying over New Zealand, Richard Pearse imagines he is moving not just through air, but over the shifting terrain of history. He drops his hat, and when he reaches to pick it up, he sees a note.

Dear RP,

it begins, and continues:

> My friend, first I'd like to say: Don't be alarmed. I am Richard Pearse, writing to you. Richard Pearse, that is, you. At this moment you are flying a plane over Wellington in the early twentieth century. Of course you know this already. (Don't crash while reading, please!) At this very moment, however, you are also in the twenty-first century, and close to you are some people dressed like they're from your time. Are you confused? The discoveries of Einstein, whom you can't have heard of yet, go some way toward explaining things, but the phenomena themselves preceded the German. To make a long story short, the activities of these people from your future, these people so interested in heaviness, have changed the topos. A new doxa reigns; the conceptual field has warped.
>
> The group in Oamaru, one of many, has created a new topology of knowledge so vivid that it is capable of affecting the 'real'. By bringing past into present with such intensity, not just through silly costumes (though these are marvelous), but through a new way of thinking, they have changed contemporary values, altering the conceptual plane from light to heavy. This has created a glitch. Past has entered present and vice versa, so that Richard Pearse, that is, you, that is, me, has turned up here. At first, I was as confused as you must be.

Then, after a while, I got used to it.

You are a writer of speculative fictions, they told me, looking at my biography on Wikipedia (meant as a synopsis of who you are available to everyone, though most are unreliable). But if you are there in the twenty-first century, who is flying this plane? you must be asking. That, I can't say. A writer of speculative fictions, perhaps. Or someone else entirely. A Maori. Arthur Eddington. How can we know?

The conceptual topos has warped, and people have begun to value heavy over light. Victorian outfits and heavy flying machines draw everything toward them, so that even when we imagine, our thoughts center round their weight. This is simply an exaggeration of previous tendencies. The lightest data fragments and letters have always been capable of holding news of weighty things. Even the fine and delicate cloth of a flag can bear symbols dense with significance, effecting even more ponderous changes...

Signed,

You

Mulberry Flag, 1872

It's time for we Maoris to take action. The Brits are attacking. We're surrounded by big colonial buildings and men with guns and we're losing, we're losing. All we have are our pā, our hill forts with defensive terraces, our walls made of pointy sticks, and these are no good against those men with their big guns. I've thought things through hard and know we need a symbol. But what? These are land wars and we cannot fight a long campaign. There have been epidemics of illness, food supplies cut off. The Brits come as full-time soldiers, but we must spend our time planting, growing. This weight in the earth, this earth we feel between our fingers, makes things hard for us.

We came together and said amongst ourselves: And what if

precisely this, our shame, ought to be the source of our pride? What if this planting in the ground of our kūmara (sweet potatoes), our taro, yam and gourd, our tī pore (Pacific cabbage tree), what if this farming is in fact what should satisfy us most? We are a peaceful people, let us not forget. Can we make this thing that weighs us down into a strength? We have to embrace our heaviness instead of feeling it a burden.

At first, we were ashamed by our disadvantage. They laughed and called us farmer boys, not soldiers, and it's true, we are not in the pay of the Crown. But we are proud. And so we made a big splendid flag, with a picture of our vegetables. Our root vegetables in the earth, the ones we rely on. Now I'm waving our flag like mad, and those poor Brits, they're stopped dead there just seeing it. The beauty of it, the craziness. The heaviness of our symbol.

The flag, she's light as silk, made from the fibers used in the roofs of our pā. We wove her out of aute (paper mulberry), but it's our heavy symbol, our vegetables, that means so much, that means everything. Those Brits look like they want to run away now. At the end of the day, they are shy like us, not fighters. Some take steps our way. But instead of lifting their Enfields, they embrace us. They cry.

Our history has been a long and strange one, but still we can laugh together, still we can live side by side. All of us are jumbled up here on the very same land, and now I watch us make peace, and weep with relief and with joy. But I want to make sure everything will be okay. My skinny arms haven't stopped waving that flag. I wave and wave it as the sky gets darker and the first torches come out, wave and wave it as the fires are lit and the meal is cooked. I'm going to wave and wave it like this, not stopping, to ripple the silk and the fabric of time. Wave it and wave it, like this, to see through the great transformation.

Jars of Preserves

saving for its own sake and the corruption of this idea /
Netherlands

c:\Documents and Settings\jess\Desktop\Downloads
File name: monsieurdebougrelon.pdf
File name: jeanlorrain.doc
File name: richterinterviewnotes.doc
File name: ennuiloop.mp3
File name: cherryjam.doc

I shut my computer, take the page from the printer and head into the kitchen.

> Cherry jam: Take 12 pounds of just ripe cherries; remove their stems and cores; add 2 pounds of currants, prepared in the same way as for the currant jam above, along with a pound of raspberry juice; put it all in the pan on the stove; boil and skim; after half an hour of boiling, add three-quarter pounds sugar per pound of juice; let boil for half an hour, remove from the heat and pour jam immediately into the pots you have prepared.

Even those who venture to prepare a meal of multiple courses only on rarest occasion can find delight in the pages of a well-written cookbook, approached as literature. There are far worse ways to pass the time than to skim the recipes of Louis-Eustache Audot's 1823 *La Cuisinière de la Campagne et de la Ville, ou nouvelle cuisine économique*. Here, a parade of jams takes the form of confitures and compotes, preserves and conserves, fruit curds and spreads, confits and marmalades, chutneys and jellies, glamorized into every luscious form of dessert possible.

The preserves, which under other circumstances might seem cozy (a spread for toast as a four o'clock pick-me-up) or sentimental (Dostoevsky stirring raspberry jam into tea with a little silver spoon) are here described in such detail they come to seem almost sensual. Certain aspects of their being start to appear sinister, transforming childishness or practicality into maturity and pleasure, oriented toward death.

Is this too large a claim for a jam? Maybe so, but the activity of preservation always holds obsession *in potentia*. It's possible to preserve almost anything: sugared fruits and vegetables in wobbly molds of aspic, cheeses, meats, paintings in museums, collected works, letters, photographs, items of clothing, memories, butterflies pinned on paper, hunted animals, human bodies.

The escalation is disturbing, because there's no clear place where you can say that this or that is too morbid, this or that isn't natural. It's easy to understand how one can derail into *la préservation pour la préservation*. How can we understand someone who saves for its own sake? And even more confusingly, how can we understand someone who pretends to save for its own sake, yet in reality uses and displays objects, corrupting the pure artistic idea of preservation?

Enter Jean Lorrain. Within the Parisian society of late nineteenth-century France, he was a self-conscious gadfly. Rather than hermit himself away at home to write diverting fantasies, he threw himself into the swing of things, finding ample material in the fancy costumes, sights and smells of Paris to capture in prose. What he did in literary terms is akin to taking the wooden spoon used for stirring domestic and worldly elements into a sugary concoction, and employing it to batter those close to him on the head. Sallying forth everyday with his lush mustache, fat be-ringed fingers, heavy-lidded eyes and puffed-up chest, Lorrain wrote criticism so scathing that men and women frequently challenged him to duels, and he lived for the rush of excitement.

In addition to his polemical journalism, Lorrain also wrote a prolific amount of narrative literature, which got him into trouble for obscenity charges. But what an odd kind of obscenity it is:

> I told you already and will repeat it: you will find the most spellbinding nostalgia in jars of fruits and vegetables ... Vegetables first, what a source of the fantastic! The old Flemish painters understood this well when they introduced into their anatomies of devils and compositions of monsters, into their Sabbaths and Temptations, all the fruits and vegetables of creation.

Monsieur de Bougrelon, the one who speaks these words in Lorrain's eponymous novel, is a personage in the fullest sense, a withered old dandy wandering about Amsterdam. He introduces strange pleasures into the lives of the ennui-inflicted men who visit him in the city, but not the sort one might expect. After an explanatory spiel, Bougrelon takes his guests to look at jars of preserved vegetables and old fabrics, to imagine women who are long gone or never were.

> The cloakroom of memory!

> In a suite of rooms lit by high windows, display case after display case lined the walls, immense armoires of glass like blocks of ice, where the fashions of lost centuries seemed frozen. Touching preserve jars of outmoded elegance, the so-called costume galleries were where the meticulous Dutch guarded the gallant castoffs, dresses, gowns, and jewelry of former queens, shielding them from the dust and humidity. Next to the long pleated robes imitated by Watteau, there were rural scenes by Pater, gros de Tours woven with silver

> fleurs-de-lis upon the Bordeaux-red backdrop of sack-back gowns, delicate striped pekins beside braids of silk, brocades of green myrtle foliage, glazed satins like rivulets of frost with Astragalus flowers and love knots, and garlands of blooming carnations and sweet alyssums tied to the fabric with ribbons ...

Perhaps this cloakroom is not so strange. One of my favorite still-life paintings of the Dutch Golden Age is Willem Claeszoon Heda's *Breakfast Table with Blackberry Pie* (1631, oil on panel). A 'breakfast piece', it preserves small moments, objects that will rot or in some other way fail to survive: transient things, fleeting reflections, glints. Here is the metal spoon perfectly prepared to break the crusty surface of a tart, to scoop glistening berries onto a plate. There is the crease in the cream-coloured tablecloth, unfolded for a special occasion perhaps, after months spent tucked away in the cabinet. Certain touches like a broken glass and a key remain enigmatic, speaking to something that happened off canvas.

Other Dutch paintings show interiors and families, and in the fresh-scrubbed faces, reflecting surfaces and glistening shells, there is both sadness and eroticism: memento mori, flesh and death, trinkets that speak not of the past but of the future as a reminder we will die. Lorrain's objects work in this way, turning backward and forward at the same time, suggesting in their decay that the past is over, yet producing these sensual past memories anew.

Preservation as conservation becomes preservation as provocation, and in so doing transforms into preservation as perversion. Usually one conserves to prevent injury, decay, waste or loss, but the sort of conservation here actually extends loss. Pickled items are kept not to be consumed, but to be frozen in their death state and dwelt on infinitely. What kind of pleasure exists in this kind of preservation? Can there also be perversities of the emotions? Monsieur Bougrelon certainly alludes to the pleasures of the flesh. In Lorrain's time, it was fashionable for artists to 'go mad', perhaps in part from syphilis, which Lorrain himself had, but mostly as a performance that entitled them to otherwise indefensible extremities of behavior. Lorrain delights in the strange thrills to be found in what is decayed, which serves as a means of accessing the past.

For Lorrain, people are still lifes, and still lifes are holders of memory, representing a longing not for something concrete but for nostalgia itself, in a double remove. It's strange to think of desire as not for a person, but for desire: a wax fruit kept on the table not to bite into, but to tap a pencil against, or juggle, or throw against a wall. Lorrain finds pleasures in the activity of remembering, the extended thought made possible by melancholy. Decades later, Proust would stay inside wrapped up for years in bandages with the lights off, turning this mummifying to glorious effect. But Lorrain wants to focus on the external preserved thing, and isn't much interested in inner life. (Is this the absence, the not-there, the […] that so much French theory decries?)

Amsterdam, the city where the story transpires, is portrayed as a place of ugly northern people and fog, Schiedammers and *Bruges-la-Morte*. Men wander about pretending to be what they are not, intrigued by the tales of other men also pretending to be what they are not. This is a preserved city, where everything is still and stopped. The wide canal is lined with houses that have little windows, squares you'd like to open like the flaps in Cadbury Advent calendars, in which the days count down to Christmas, elf-shaped chocolates waiting.

Interiors are where things happen, rooms full of revulsive old things, withered hands, faded fabrics, dolls and dolled-up girls with no souls, cadavers and 'lavish historians' tending the works. In this environment of infinite sadness, a tender compassion is cultivated, the rot already in the apple. This is a brothel not of bodies but of pasts: a space dedicated to the decadent pleasures of living in one's head, in a kingdom of eternal memory. Nothing magical here, just regrets and thoughts, someone says at one point. Coldness, ennui, preserves. During the writing of this essay, I opened the refrigerator and looked carefully at a jar of pickles. It had a gold lid and inside, floating in the vinegar, were baby sour gherkins. Cornichons, product of France. They were delicate and green and a little repulsive, and they continued to absorb their spices as they underwent the turbulent journey within the jar in my hand to the counter. A slow drift, then they were settling back into position at the bottom of the glass cylinder with grace, in gentle collision.

Almost nuzzling, the way you can imagine that dogs or other small animals do. Within the vinegar, the pickles would never be able to move very quickly, a limit on their freedom that would perhaps always exceed their comprehension; though one can never really be sure about these things. I was going to eat one of the pickles with a bit of cheese, but honestly, after considering them so long, I found I'd lost my appetite and just put the jar back in the fridge.

Ennui, the state that permits an obsessive attention to detail, which Monsieur de Bougrelon encourages and finds artificial means to bring about, is not the laziness of lying in bed, but rather a semi-humorous, semi-serious way of drifting, looking closely. How did Lorrain come to write his books? Was it the ether he consumed? Perhaps that too, but he wrote so many with such a specific preoccupation that his attraction to such aspects of the melancholic and derelict must also have just been a part of his personality. Imaginary pleasures, bodices of nothing, cloakrooms of memory, jars of preserves, a whole philosophy of the still and quivering. It isn't easy to square Lorrain's reflections on sadness, objects and dead things with his sharp public personality as a dandy, always searching for the whiplash bon mot. Or perhaps it's all too easy, since sharp writing about trivial things is one of the greatest of decadent pleasures. Perhaps harder to understand than the motives of the writer are those of the reader.

In her book *The Art of Describing*, Svetlana Alpers distinguishes between the textual imagination of the Italian Renaissance and the visual imagination of the Dutch Golden Age. The visual imagination requires things to be still, to describe them. But to the point of deadness? Why look at a still life at all? What is the attraction of something as quiet and unmoving as a *naturaleza muerta*? Perhaps this is the same question as: What makes a person want to watch the slow performance of a tragedy? For the catharsis of finding joy in the trigger of old memories? Why do people purposely seek out experiences of tragedy and objects that provoke such a state?

Monsieur de Bougrelon was originally printed in 1897 by Librairie Borel. The release of two translations of the book now, along with the start of several presses over the past decade devoted to the period, makes me curious about why there's been a resurgence of interest in this kind of book. Perhaps it's just a coincidence of editors coming along now, at this point in time. Or perhaps it speaks to a larger cultural moment. Maybe there is a place for a decadence of thought in which nothing actually happens. 'Oh the exasperating mystery of the sorbet that never melted,' says one of the characters, talking about a woman.

One of the two translations is by respected English translator Brian Stableford for Side Real Press in the UK, the other by Californian translator Eva Richter for her own editorial house. I chatted about the book with Eva, who lives in Palm Springs, a place at first sight in all ways opposite brooding late nineteenth-century Amsterdam. The dark underbelly and seediness concealed behind the sunshine of the place are precisely what interest her though, driving both her writing and the selections of her publishing project Spurl: a nonsense word, loosely modeled after the name of Vienna coffee shop Café Sperl but deliberately Dadaesque.

One of the last few editions Eva put out was Barbara Payton's *I Am Not Ashamed,* the memoir of a blonde actress who appears with furs and a defiant look on the cover and talks about her past dealings with men. Another was the excellent Swiss writer Henri Roorda's *My Suicide.* (I'd like to see someone translate his *Le rire et les rieurs*.) The editions are simple and unprepossessing, the contents striking enough. 'I want to do something interesting and appealing, but also boundary pushing,' said Eva. Her bookcase behind her, sheltering hundreds of never-translated volumes in French, suggests that this might become a life task, or at least a limitless diversion.

Imagine going about life seeing it as possible material to be preserved. Every tiny quiver of a blade of grass, every shine or shadow, every rustle, would become an optical or aural challenge to 'keep'. What would stay fresh, what would spoil and go rotten? Can one create a private archive, in which it is possible to save things and revisit them in solitude, away from the world? How should one think about memory, how should one cultivate it? As a personal museum somewhere between sacred chapel and Mnemosyne Atlas, as a rosary for remembrance, as an eternal jar of pickles, as an Advent calendar that endlessly restocks the sweets behind its flaps, so eagerly torn open?

Different reasons exist to preserve things. They can be preserved because they are useful and later meant to be consumed, like a jam. They can be preserved for display, like the objects in a Dutch Golden Age painting, which show what a family owns and values.

And they can be preserved like Monsieur de Bougrelon's objects, intended for both display and use. The physical apparatus of modern computers works like Bougrelon's cloaks, in that it is meant to be both aesthetic and useful. What is inside the machine is different, more like a cherry jam meant for pure use, files saved to memory in the binary code of 0 / 1.

Here, again, we come to the idea of preservation for preservation's sake, 'just in case'. Items are 'autosaved', stored away, whether or not they will be used later. Data is saved not to consume like a jam or show off like a Dutch painting, but 'just because' and 'in case'. Maybe we will open a document later, maybe not. The things retained can potentially be opened again and revisited, the same memories relived multiple times. Just as Monsieur de Bougrelon's physical object evokes an invisible person, a visible document evokes an invisible memory bank of data encoded in binary numbers, in computing terms 'persistence'.

There are interesting consequences to the machine's saving for its own sake. The preservation of silly, non sequitur things, for instance: items that quickly expire, pomp on behalf of nothing, material for *Spontaneity: A History in 12 Volumes*. Once attention has been drawn to the process of saving, there are all sorts of ways to play with the concept.

Yet, here, an updated opportunity is also created for those who take the idea of 'saving for its own sake', and give it a further turn of the screw, making claims on behalf of the purity of saving, but in reality displaying to show off, and using what is supposedly there 'in (display) case'. Rather than embrace silliness our waxy, wily Monsieur de Bougrelon saw opportunity and swung his walking stick; using the framework of pure art and preservation for preservation's sake, and moving in multiple and subtle ways to corrupt it.

The Decay of Decay — a Drollerie

One day an epidemic began of what appeared to be rot, but was not so. The people thought that what they saw was decay, genuine rot, and filed the appropriate complaints. In reality, what they suffered from was false decay, an appearance of rot that coated everything like gilt on a box. False, yes. The decay had been put on the canal boats by the Preservation Commission, an organization set up to preserve old things in the city. It did so, but also took steps to add 'authenticity', touches of fake rot. After some time, those living in the city started to get used to this. They withdrew their complaints, and spoke to visitors about the long history of the place where they lived. The conservation commission devoted itself to full-time preparation of the fake decay, to avoid any freshness. Complaints began to be heard whenever a colour was too bright or a deck too well scrubbed. Real decay became unpopular. Nobody wanted green or brown; they preferred the beautiful, specially prepared verdigris of the Preservation Commission. Only one woman, more out of boredom than anything else, took the trouble to save a few samples of original rot. A bit of rust and cracked paint on her deck was tucked away in an old powder compact. Soon afterward, the authorities knocked on the door to replace what was in her backyard with a better kind of ruin.

Brandy Cocktail

inventing styles and pseudonyms to avoid spectacle / Bolivia (reprise)

Yard upon yard of fabric for elegant costumes, extra upon extra for inclusion in deluxe crowd shots, pound upon pound of gold for the sensuous glide of the opening credits. The impossibly well-funded costume drama *The Crown* aims to enthrall viewers in the UK and US with its depiction of the royal family, and on the whole it succeeds in doing so. Along with haute couture, 1930s banter is on display. This was a decade in which Hollywood screwball comedies with zippy exchanges gave spice to romance, and even in the highest reaches of UK government, where upper lips seemed stiffly starched, verbal fireworks were an everyday affair. Such, anyway, is the impression one gleans from the show.

In real life, to supplement whatever natural wit might have been 'in the air', comedy professionals were often invited to Buckingham Palace. One of King George's favorite groups was the Crazy Gang, a highly choreographed slapstick act. Two members of the Gang, the real-life married couple Billy Caryll and Hilda Mundy, also recorded side programs for radio, including a sketch called *Scenes of Domestic Bliss*. In quick back-and-forth dialog, they quarrel without humor but with rhythm. Here, perhaps, is what the queen's subjects got up to while she reigned.

The recording begins with Hilda singing and Billy accusing her of making a sound like killing a cat. 'Shut up,' she says to him. She wants him to eat his breakfast, and tells him to stop drinking. He tells her to go to the doctor, since she's suffering from chronic complaint. She says she's cooked eggs. 'Eggs again?' he moans. 'It's always Easter in this house.' The marriage jokes begin. 'I cook and cook and what do I get for it? Nothing.' 'You're lucky. I get indigestion.' 'A few words mumbled over your head and you're married.' 'A few words mumbled in your sleep and you're divorced.' 'Clever men make good husbands.' 'Clever men don't become husbands, they're too clever.' He makes fun of her mother's teeth. She sends him off to work.

Later, he comes home very drunk. 'Where have you been for the last four hours?' 'Talking to a barmaid.' 'And what did she say?' 'No.' 'If I were in your condition I'd shoot myself.' 'If you were in my condition you'd miss. Let me go to bed, you wicked woman, or I'll apply for Restitution of Congenial Nights.' (An allusion to the Restitution of Conjugal Rights Act valid at the time, abolished in 1970.) 'You're not a naughty boy, you're an imbecile is what you are. And I'll tell you something else. You're a nasty, drunken little beast. A beast, I tell you, a beast. I hate men.' 'I hate women. And I wish there were more women, so I could hate them, too.' 'You wicked brute. Where do you expect to go to when you die?' 'I don't care. I've got pals in both places.' She gives him back her wedding ring. He leaves, slamming the door.

The recording is from 1934 and very much of its period, the eternal tension of the couple relying on the premise of a man who does whatever he likes and a somewhat shrewish wife who devotes her life to tracking him down. There is a pride in their bickering, a pleasure in the jabbing. Marital chaos becomes material for ostentatious show.

Scenes of Domestic Bliss is set firmly in 1930s Britain, but it didn't stay there. Somehow, word of Hilda arrived in the town of Oruro in Bolivia, famous not for crowns but for miners' hats, as one can see from the city's enormous monument. Located in the west of the country just south of La Paz and Cochabamba, today the city is mostly known for its carnival celebrating the Virgen del Socavón, or Virgin of the Tunnel, a useful deity for the workers underground.

Since Hilda hadn't made any of her movies yet, it must have been through a recording like *Domestic Bliss*, possibly brought home by her diplomat father after travels through Europe, that Laura Villanueva Rocabado encountered the woman whose name she would assume. Throughout her life Villanueva would use a variety of noms de plume (Raspadilla, Retna Dumila, María [Motia] Daguileff, Madame Adrienne, Jeanette, Laury, Pimpette, Dina Merluza, Michelin, Mademoiselle Touchet, Dora Kolonday, Ana Massina), never signing as her 'real' self. That Hilda Mundy is the name under which her major work *Pirotecnia* is edited suggests a certain priority, however, and it is under this name that contemporary publishers have decided to group her work.

Thanks to *La Mariposa Mundial* editor Rodolfo Ortiz in La Paz, readers now have access to a splendidly organized new edition of Hilda's journalistic work called *Bambolla bambolla*. The text includes a timeline, maps and photographs, and it begins with a series of letters exchanged with another writer, who compliments her on her spark. Hilda began writing and collaborating with contemporaries in Oruro, and continued to write in La Paz. Her column titles consistently have interesting names, such as 'Dum Dum', 'Vitamins' or 'Brandy Cocktail', reflecting the chaotic influx of foreign publications from the UK and US on the Bolivian middle and upper-middle class during this period.

Dum Dum was the name of a publication Hilda wrote with her friends, which came out on Sundays. Reading the phrase Dum Dum, I thought immediately of the lollipop, available in all the colours of the rainbow and popular with the kids at my elementary school; the first factory was founded in 1924 in Ohio, so it could have been an influence on Hilda.

Ortiz convinced me that the name actually probably comes from the British Dum Dum, a kind of exploding bullet produced in a factory in West Bengal, India. There is also a 'Dum Dum' song by Brenda Lee and a contemporary Brazilian rapper with that name, fascinating both but anachronistic as influences on Hilda. Given the roving curiosity and avant-garde aesthetic of Hilda, however, nothing would be surprising.

The real Hilda is a mystery. She plays with absurd arguments under various names for the sheer pleasure of it, and often takes a tone of exaggerated umbrage over some annoyance of daily life. To read Hilda is to discover the small, the fragmented, the specific, the hands-on. (Literally so: later texts of hers would show up in the Argentine publication *Manos maravillosas* [Marvelous hands], devoted to artisanal craft.) In an essay about her work for the newspaper *Página Siete,* Rocío Zavala Virreira writes that 'to speak of Hilda Mundy is to leave the path, change direction, try out new things. It is to think not in terms of books, but magazines. Not complete sets, but clippings or incomplete collections. Not fire, but fireworks. Not war, but bullet shards. Not a big picture, but impressions. Not a current, but a short-circuit. Not alcohol, but a brandy cocktail.'

Hilda's work was not collected until in the last few years the efforts of La Mariposa Mundial and Chilean publisher La Mujer Rota began to gather her work into book form. What does the title of Hilda's book mean? *Bambolla* means something like show or ostentation, so *Bambolla bambolla* suggests a comedic 'look at me, look at me'. The expression is a phrase from the poet Góngora, which 'Laury' paraphrases in a letter. '*Bambolla. Bambolla. Ande yo caliente y ríase la gente,*' she writes, tongue in cheek. ('Showing off. Showing off. I go about worked up and people laugh.') The title leads one to think about what the relationship between ostentation and art might have been for Hilda.

In English, 'ostentation' has gradually gone from meaning *an act* of display (in Catholic context, a presentation of being, as with the wounds of Christ's wounds in the Passion) to *an excessive act* of display, a transformation of 'show' into 'show off'. As every Dum Dum-holding third grader too clever for her own good knows, a group of peacocks is called an ostentation, too. I don't know the genealogy of *bambolla,* but this double idea of show and show off is useful for thinking about how Hilda conceives of display.

For all she liked jokes and assuming different social identities, *bambolla* for Hilda was not necessarily comic. In wartime, for instance, it simply became ridiculous, and war presented new forms of spectacle. Hilda's 'Impresiones de la Guerra del Chaco', one of the texts in *Bambolla Bambolla,* is both a historical document and a literary work of what might be called journalistic poetry. Running from 18 June 1932 to 17 June 1935, the impressions follow Hilda's thoughts as a spectator in Oruro to events in the Chaco, when soldiers from Bolivia and Paraguay set about destroying one another at great economic and moral cost to both nations.

Hilda tracks changes in language: 'news-sparks' of developments detonate in the events that follow. The 'I' of Hilda and her spirit interact with the disembodied voices of those around her, the craziness of society, the hostility of nature, the mindless brutality of machinery. Her fragmented impressions of the Chaco War, which later commentators would refer to as an *infierno verde* (green hell), are a personal register of events as they occur, in spiky poetic prose.

Despite their on-the-fly character, the impressions are conscious of being a text. Hilda had already written the texts that would form part of *Pirotecnia,* and her notes display the avant-garde techniques honed in those works, written in bursts like quick clips of machine gun fire. Across the world in Italy, Marinetti's Futurism was trying out its legs, and in later texts Hilda would discuss futurism directly. Here, its forms are prefigured. But this does not mean there was not a romantic element. Hilda wants to outline scenes in disconnected, spontaneous notes, not 'sterilize' her style. She says Fra Angelico did the same, painting angels just as they came to him, trusting even their imperfections were meant to emerge as they were. The first words to come forth were the ones that reflected truth and spirit.

Head in the clouds as this may seem, Hilda's verbal shrapnel is anything but too soft. At the start is a discussion of what the 'retinas' that look at her phrases will find. (Scientific terminology is everywhere.) There won't be 'beauties of style, rigid histories or waves of meditative philosophical phrases', just text from a sensitive spirit that took in war 'as easily as ice cream'. The experience of war is no different for Hilda than any other form of experience. In fact, it transmits almost too cleanly as spectacle, with 'landscapes of war' just another kind of painting. Hilda says she sees no need to read newspapers and only unfolds them to make paper birds. In reality, from a young age she reads the newspapers and talks about her impressions of events with her uncle and father; although, she is very conscious that the pages are fragile, and dubious as vehicles of truth.

Rumors of war start to appear two years before war itself, brought by the 'shadow of a bewitching and fatal destiny'. Hints of conflict are present in the air, the sun, in life itself. The days leading up to war are drawn out, tortuous. 'Abstracted in fixed visions, my fingers stiff with cold, I wrote war stories, morbid fantasies, crucified loves. My imagination marked out exactly what I lived afterward,' she writes. News of the start of official engagement arrives on a winter day during the glamour of the inauguration of the Feria Nacional in La Paz. This festival brings out a 'mad enthusiasm' strangely similar to that of the war that follows, and Hilda is attentive to similarities in the preparations and festive atmosphere.

Bolivia and Paraguay go head to head, and people stream over the chessboard of city streets. The city is rational, the people are not. The 'monster of the collective' demands blood. 'Blood', shouted loudly and printed in newspapers, is converted from text into violence, as the verbal becomes physical, a terrifying act of transmutation. A single word repeated over and over draws the atavism of crime from thousands of souls in the name of patriotism. Without irony, Hilda refers to her 'feminine heart' and 'intuition' as being able to perceive the cruelties and absurdities of wartime developments. She looks at the military technology, the trains, the long evacuation convoys of the sick, the hurt, the mad. The military men are pale, afraid; their soldierly mystique has long disappeared.

Hilda's hometown of Oruro becomes a railway center, and every few hours military convoys come through, full of soldiers 'drunk with enthusiasm'. Meanwhile, evacuation convoys return in the other direction. Strong men leave; weak and moaning men return. As a spectator one step removed, Hilda sees how futile it is. In a fascinating editorial intervention by Ortiz, two photographs are shown side by side. The first shows soldiers standing in a neat line, the second a column of what Hilda called 'infinite caravans' photographed by her brother, a chilling similarity.

At one point Hilda summarizes three years in a single night, and says the details have fled her recollection. 'My spirit sick with neurasthenia and memories relives the past war as if it were a nightmare.' She works as a typist by day, writes by night. War itself always remains just outside the frame, but Oruro is a sensitive register of changes. People get used to living in the 'shadow of the Apocalypse', in constant crisis, and cheer every development. Real or imaginary, there appear in Hilda's notebook last messages scrawled by those heading to war: 'I leave for the training camp with a secret fear that casts a veil over my horizon, now farther off than ever before. When you read this, remember our schooldays. And if you hear something fatal, send up a prayer for my soul,' one says.

The restlessness of Hilda's eyes continues. She speaks of her emotions, the petty bourgeoisie and masses, the criminal kings and soldiers, the 'tiger race', the way people make war a business opportunity. In one entry she turns to lyrical poetry, and the wind becomes a symphony of sad motifs. 'I wandered far and near,' she writes. 'The wind, naughty thing, brought me a slip of paper. My Eve was startled into curiosity. I gave way and looked at the paper.' She sees a list of expensive food items for purchase, a sign some are living in decadence at a time of frugality. On the same page in the book, there is a picture of her two brothers and three other boys she knows from student days, all soldiers in the Chaco. Speculations in high finance, an operational black market and a leftist opposition all thrive, created by war.

Hilda writes these notes to bring information to life and give it blood. Official history is only for nationalist use, teaching one to cheer or lament in the right places, she says. Sometimes she gets angry and uses capital letters, like a page from a pamphlet. She refers to the *etapa-fárrago de nuestra Historia* (hotchpotch period of our History) and gets frustrated with small-time traders as well as those from her own background. It's annoying to see how gullible the middle and upper classes are in their *zarabanda loca* (crazy whirling around), an attempt to get involved.

She is also perceptive about the self-serving political positions taken by other nations: 'The international "camouflage" of the Argentine nation inspires mistrust in us. Every time it intervened, it was to mediate with damaging clauses. In contrast, hope smiled at us from Brazil.' She talks about candidness and masks, about how the whole war has come to seem a piece of yellow journalism fit for a dramatic image by Dürer. The text ends abruptly with a 'final chime', no clean summing up, reflecting her lack of interest in building arguments.

During the Guerra del Chaco, developments in military technology changed the battlefield in unimaginable ways. Just as the introduction of the tank made an enormous impact during the First World War in Europe, tanks now appeared for the first time in South America. A number of smaller new technologies were also tried out for the first time. Hilda was generally enthusiastic about technology. She typed up her pieces on a Royal typewriter, and rhapsodized over the possibilities of the telephone and electric tram in *Pyrotechnics*. But the technology associated with war was a different matter.

Military tanks, railroads, parades and photography are all forms of ostentation. Amusing in some contexts, in this case display took on a sinister quality, as it was in opposition to traits Hilda referred to as 'intuition', 'spirit', even 'God', a 'celestial and Divine' within ourselves not associated with organized religion. The ostentation of the parades sickened her. 'In this state of things, nice and prepared for the act, brave young girls and buxom ladies saw and felt enter into action all the machines of war: 75 and 105 tanks; mortars of heavy and regular capacity; machine guns, even hand grenades …' Hilda's response to this equipment and self-congratulation was a kind of hermeticism, reflected in her notational style. Instead of a comforting lushness of detail, she opted for hard, compact, suggestive prose.

Ostentation can take many forms, from fancy cocktails to British costume drama coronations, military parades to theatrical 1930s-style humor. Sometimes this excess and display can be a source of pleasure. Other times, stylized presentation can invoke unease and illness in the spectator, making one ask what something is trying to achieve. What lies behind the compulsion to write? To show instead of stay silent? Where is the line between showing and showing too much, between necessity and spectacle? How is surface related to depth?

Maybe Hilda had answers to these questions, maybe she didn't. Maybe her life turned to other priorities. Maybe she kept writing just as many sharp and witty texts at the end of her life as at the beginning, but didn't feel the need to send them to the papers. The amount of Hilda's published writing diminishes massively after her first book *Pyrotechnics*, around the time she married her husband. But the relationship between Hilda's staccato imaginative production and her real-life 'scenes of domestic bliss' is a kind of tease. It's quite possible this was her own choice, linked to a Dada aesthetic of choosing life over the 'show' of writing. Everything she wrote took the form of sketches, but deliberately so; as if this itself was a challenge to those who would make art into parade or glossy display, an overly technical 1:1 model for life rather than something more allusive.

When the queen removes her crown, she is an exhausted, uncertain woman. When the vinyl stops spinning, Hilda Mundy and Billy Caryll speak or quarrel off the record. Far from the parades in La Paz and Oruro, war in the Chaco slogs away, unseen. When Laura Villanueva Rocabado closes her book of notes, she goes on beyond reach of her readers, free to live her life.

Decision

Write a poem
to cap this essay?
No, I'll fix myself a nightcap.

Symbolic Grenade

observing as a mode of history and form of action / Pyrenees

A unified theory: that was the dream of Cambridge mathematician Srinivasa Ramanujan. A correspondence between the marks on paper and the structure of the skies, a single formula that could encapsulate the universe. Ramanujan wasn't successful in finding his theory of everything, but his ideal remains a covert presence and source of strength for many who have embraced a life in the sciences. But what happens when that ideal is applied to history? Is history a science, too? Is the attempt to create a blueprint misguided if we're talking about human endeavor? Or can one look for a pattern there, as well? If so, how should one go about trying to find it? Is it best to remove oneself from the world to ensure peace of mind and the tranquility necessary for tracing larger arcs? Or should one try to be as actively engaged in daily life as possible? Do the aims of history-writing undergo development, in the same way that ideas of modernism marked a literary shift, partly in response to scientific discoveries? And is there some shining pattern or arch-truth behind these changes? Or is history just an infinite parade of possible anecdotes to arrange, catalogue, exhibit, assemble and frame in a Duchampian exercise, like a box of old film reels? Can the historian in his observational role play some part in affairs, creating change through his attempt to understand? Or is this withdrawal into the imagination folly?

This flood of questions lies behind Carlos Fonseca's intelligent and elegantly written novel *Colonel Lágrimas,* translated from the Spanish by Megan McDowell (Restless Books, 2016). The book takes the form of the notes of Alexander Grothendieck, a mathematician who retreats to a remote location in the Pyrenees and embarks on a tragicomic project to discover an allegory for the political trajectory of history.

Dissolving into his story, transforming into an enigmatic presence that comes to seem like time itself, Grothendieck refers to his own persona throughout as the 'colonel', possibly owing to the efficient, quasi-military way he goes about his writing, possibly because of how he views the past as a battlefield of meanings. Fueled by alcohol and coffee — after all, ours has been a 'caffeinated age' — he sets about his work.

Left alone for hours with his texts, the colonel jumps from one volume to the next, 'making impossible connections between impossible continents and eras'. *Vertigos of the Century,* his masterpiece in progress, is intended to be a blueprint of himself and all the events that have occurred in the last hundred years. He is obsessed by the beauty of brevity, the elegance of simplicity, and he wants to fit everything that has happened into one equation. Abstract symbols become a way to make sense from a mess of meaningless facts. And so he takes notes for an 'autobiography of others that gradually extends over everything, threatening to become infinite'. Jumping from one idea to the next, his prose fragmented and structurally demented, he writes and writes, yet nothing actually happens, or everything.

Is he a genius or is he cracked? In the cover image, the colonel's head explodes in a scrawl of black ink, as he holds an enormous pencil to mark out his lines on graph paper. The image may just have been a publisher's choice, but the colonel doesn't really come off as deranged; just an exaggeration of the tendencies many introverted or solitary people exhibit. Obsession with some big idea combines with close attention to detail and a wandering mind prone to distraction, whether the source be drawings in the shape of Cuban 8 loops or a Jacques Brell chanson.

There are dangers to an existence so entirely oriented toward thought, of course. Seen from the outside, the author is a man who is 83 years old, schizophrenic, in dialogue with the voices in his head. In an absolute 'loneliness populated by ghosts', he analyses and reanalyses the same events from infinite angles. Guilt is a major theme, and to some extent his project is also 'the equation of his guilt'. His life is the story of the things he didn't do but should have done, and he alludes to an old lover named Cayetana Buamante. 'Was it so difficult to take a plane, load up the chalkboards and some chalk, and go and build the revolution in America, that continent that saw his birth? So difficult to follow the footsteps of that woman he loved and of whom he now only has some photographs without memories left, a few lifeless equations?' Considered this way, it seems easy to criticize excess passiveness as a negative historical position, and glamorize action and living in the present moment.

'Not understanding is perhaps his way of life,' the colonel writes. 'This same man who, as a child, would distract himself from hunger by untangling invisible knots, this man whose passion gets tangled up among doodles and equations, knows full well there is a beauty that consists of allowing oneself to be caught up in life and following it to its final consequences.' And so, battling amnesia, he sets about trying 'to codify life in small postcards, build an encyclopedic Babel for his cracked memory' and begin to reconstruct his past.

The reiterated question becomes that of the relationship between mind and action, which is both a historian-actor problem and an observer-participant problem. Do one's 'public persona' and social activities matter, or is what matters the life of the mind? When the colonel is photographed, there's an artificial quality to it, and he 'seems to assume the posture of a dandy posing for the posterity of a futuristic camera'. What is the line between truth and the public appearance of truth? And at what point do impressions cease to matter and give way to the story told about them afterward, reducing people to their appearances? At one point the colonel writes a postcard to his character Maximiliano:

> You know, Maximiliano, that this Ronald Reagan, man of a thousand facets and a dapper walk, illustrious president of the United States, had the most interesting job before he found success as an actor: he was an announcer for American football games. The strange thing, the magnificent thing, Maximiliano — and here is the point of this anecdote — is that this future president didn't watch what he was narrating: he simply received bits of information, strung like rosary beads, whose whole he never saw, loose bits of information about a spectacle he didn't see, but whose tone he imagined in a kind of blind broadcasting. Our project is a bit like that. Broadcasting for an age without witnesses, a kind of blind narration of this dance of crazies. So, learn to tell without seeing.

An obvious displacement exists everywhere, between mind and behavior, event and interpretation, fact and memory. A constant slight sense of slippage suggests the colonel has not quite been in the right place at the right time. 'At times we feel that *Vertigos of the Century* is a kind of blueprint of the colonel's true heroic attempts,' writes the narrator. 'At times we feel that our hero arrived late to the epic of his age. He was there — in the Mexico of the twenties, in the Spanish Civil War, in the Second World War, at Woodstock and Vietnam — but always a little before or a little after, a little out of time and place.'

Although he is the only character, in a sense the colonel is still in good company. He imagines a cast of other beings, his 'divas'. These include a female artist named Chana Abramov, an anarchist named Vladimir Vostokov battling technological modernity, and a man named Maximiliano Cienfuegos (recalling the monarch of the Second Mexican Empire Maximilian I and the Cuban revolutionary Camilo Cienfuegos), who gets involved in revolutionary political activity. For the colonel, reason is not the opposite of imagination; it is what gives fuel to his invented images. His 'great work' is mirrored in the minor endeavors of various characters, such as Maximiliano's 'Diatribe Against Useful Efforts: A Thesis Against Work in the Practical Age', critiquing useful work. The colonel himself maintaining pages of notes that link seemingly disparate concepts, such as 'an entry about tightropes and history, about risks and politics'.

If a grenade is thrown, it will only ever be symbolic. Anecdotes of retreat transform into anecdotes of obsession, as in Chana's obsessive and repetitive painting of:

> ... the same landscape of a volcano that little by little, with the fading of memory, became simpler and simpler, less figurative and more abstract, and that went from being the recognizable volcano of her Mexican years to a collection of sparse green dots on a terribly white canvas, in a kind of catastrophe of the image from which only the colour was salvaged in a posthumous ruin.

Reading is a way to remain silent, and here silence is preferred over action. The colonel writes:

> One must carry entropy to the edge of the possible, play with absolute equilibrium, and only then finally act: entering the panorama with a totalizing gesture. They are wrong, those who believe that a man who speaks is more worthy than the one who stays silent. It is a simple matter of not really doing anything, of concentrating one's energies for a future moment in which the being is fully expressed.

The spiraling involutions of the colonel's thoughts are drawn with great complexity, as well as sympathy. Action is difficult, but so is thinking, and mental life is seen to be very rich. The colonel finds delight in unexpected new forms, such as the 'pleasure of bewilderment' and 'transverse pleasures', which connect elements across time. The distancing method is appropriate for Fonseca's mission to transcend local references and do something other than simply describe reality. The colonel wants to think in a way that isn't limited to a single place or even time, as the small lives of humans are part of much bigger processes:

> The real work happens on a different plane and a different scale. Take genetics: Man takes pains to sculpt his health, his face, his body on a human scale, though he has a legacy from long before chiseled by millions of years ... The true Copernican Revolution would consist of our realizing that human ability can do little against nature ... I've said it to Maximiliano: life is not made to the scale of man, but rather to the scale of these white mountains.

The concerns mirror those of their young author. Fonseca, who published the original novel in Spanish last year with Anagrama, was born in Puerto Rico and grew up in Costa Rica. He studied at Princeton and is now a Postdoctoral Fellow at the University of Cambridge. The alternate ambition and futility of the mathematician's 'project' at times resembles a doctoral thesis. Fonseca's book very much reflects the modern(-ist) condition of a jet-lagged, on-the-go, post-national and post-temporal world, in which one could theoretically travel anywhere, even if one would prefer to stay in one place and think deeply into a single subject. 'Intelligence', always a nebulous concept, here comes to mean the ability to forge links or connect disparate concepts into a narrative.

Displacement is the situation of the writer, who tends to note or interpret rather than act, and this story of a man who isolates himself on a mountain with his memories is an analogue to the trajectory of not just twentieth-century history, but also twentieth-century literature. The lines between 'Latin American' and 'non-Latin American' literature have broken down in meaningful ways, and the question of whether or not identity based on geographical location even makes sense is a serious one. Writers like Fonseca, living abroad and writing with influences from US, UK and French literature, clearly have a sense of identity that goes beyond the place where they were born. This can often push a text toward abstraction and wordplay, not only making the result a fascinating linguistic Molotov cocktail, but also generating a literary self-anxiety, a concern with being an observer outside of events.

Observation can be fun, in some circumstances. Two of the most fascinating and magical childhood toys are the stereogram (usually called the 'Magic Eye') and the kaleidoscope. Both play with perception, with a key difference. With the stereogram, you have to stare at a repeating field of patterns until, in a brilliant moment of intuition, a 3D-shape reveals itself to you. With the kaleidoscope, on the other hand, you look into the eyehole of a mirror-filled cylinder, which you turn. Images are likewise revealed, but in succession, one after the other, never repeating. These two toys might be used to compare the trajectories of Fonseca and his teacher, the Argentine writer Ricardo Piglia, author of *Respiración artificial, Prisión perpetua, Plata quemada, La ciudad ausente* and other classics of contemporary literature.

In 2015, Fonseca wrote an essay on Piglia's *Antología personal* for the Mexican magazine *Horizontal*, arguing that for Piglia, reading is a utopic activity and political gesture. The world offers up a certain image, and reading gives it a new form. Everything that occurs seems random, but is really a code, and reading is a form of cryptonomics, a way of deciphering it. Piglia, he notes, compares his own work to a stereogram, which can reveal an unknown dimension. Through literature, the world is seen with not an innocent but a playful and mischievous eye, and this dimension is potentially political, as 'seeing' differently can disrupt the causal logic of pragmatic capitalism and draw out a sense that is not common sense.

In a way, Piglia's work responds to the challenge of social realism, with its demand that literature reflect reality. His writing argued just the opposite; that literature should operate on reality. A recurring image in his work is his personal rewriting of Borges' Aleph, in the project of 'a photographer who from his house in the neighborhood of Flores, dedicates his days to imagining an impossible project: a replica so exact of the city of Buenos Aires that, far from being the mere representation of urban life, it becomes its secret cause'. The miniature world seeks to alter the real world; the stereogrammed image can alter reality.

The opposition between social realism and metafiction was a real one when Piglia began writing. Now, though, it seems the literary landscape of even what we call metafiction has segmented, with the line between fiction and non-fiction breaking down. Some writers pursue metafiction by making their own lives into essays with an unstable narrator or altered details, drawing on but subverting the realist approach of using real life as the material for fiction. Fonseca takes the opposite tack, still working within metafiction but drawing from the Piglia- (and Borges-) inspired position that literature can alter life, even if this position has by now also become self-conscious and uneasy with itself.

What happens when life-inspired fiction goes on to alter life, for better or worse? A pseudonym or altered version of oneself is no longer avant-garde or scandalous, and is not necessarily an angled way into a single true identity, but the creation of an entirely different possible self. Without the belief the image produced by playful new self-creations will lead to a definitive image, the stereogram can become a kaleidoscope. 'Verbal kaleidoscope' is a phrase found in Piglia's blurb on the back of Fonseca's book. Like all quasi-marketing material, this image is not perhaps meant to be read with undue seriousness. But it offers a way into understanding how Fonseca departs from his teacher, as well as the challenge faced by a new generation.

While a stereogram is a way of offering a perspective, at least it retains a certain solidity, a single solution. A kaleidoscope, on the other hand, is infinitely unreliable. Words turn ever inward, yet the meaning they seek remains forever elusive. Will the colonel's 'divas' help him discover the one true formula? Or will these identities simply proliferate meaninglessly? If a faith in political commitment (or something else) is missing, what 'Magic Eye' image can one hope will emerge?

This is also a question for Fonseca and others: At what point does a multiplication of anecdotes transform into the unified vision of a book? Does this have to do with the sequencing and feedback loops of how it is assembled? I remember something Gabriel Josipovici said in an interview once, talking about *Infinity, The Story of a Moment*: 'I at any rate dream of making a work that is like some complicated toy you can dismantle and put together again, and that is always not just more than the sum of its parts, but in a different dimension.'

Whether out of the necessity of his position living abroad or his personality, Fonseca has turned to this particular way of writing, and in doing so is discovering a new strangeness and complexity to history. *Colonel Lágrimas* suggests that perhaps there is a way to understand history that distances itself from direct representation of the past to better understand it, capturing how memory and imagination saturate the texture of the historical fabric in complicated ways.

Fonseca's Cambridge PhD dissertation, 'States of Nature: Catastrophe, History and the Reconstruction of Spanish America', explored the emergence of catastrophic representations of history. This was within the transatlantic historiography of the so-called 'Age of Revolutions', which marked the collapse of the Spanish Empire, as well as the subsequent foundation of the Latin American nation-states. One might similarly think of the behavior of a kaleidoscope as that of a disaster. One image becomes another, and then another, and then another, unpredictably. There is no state of nature, only infinite change.

In this situation, words become even more important, for as in situations of unstable reality, the most convincing idea wins and shapes the reality around it. In the phrase of J.L. Austin (an Oxford man), the question becomes 'how to do things with words', how to use language to shape actions in the world. Literature doesn't just reflect reality, but can operate on it in indirect but real ways.

At some point we get a glimpse of the colonel's shelf: a Mexican translation of *A Thousand and One Nights*, the Islamic eschatology of *Kitab-al-Miraj*, Shakespeare's *Tempest*, Cicero's *De Inventione*, Quintiliano's *Institutio Oratorio*, Mao Zedong, Simon Bolívar,. Joris-Karl Huysman's *À rebours*, an edition of *Tristam Shandy*, and Flaubert's *Bouvard and Pécuchet* are all there, among others. These influences inform Fonseca's book, which in various places also seems his personal homage to other writers, from Thomas Bernhard (*Correction*), to Gabriel García Márquez (*No One Writes to the Colonel*), to the lepidopterist Vladimir Nabokov (there's a description of the colonel in short pants in the middle of a Swiss river, holding a butterfly net to catch specimens for arrangement in his private collection). There's a dark comic element too, along the lines of some works by António Lobo Antunes, or Joaquim Maria Machado de Assis in *Posthumous Memoirs of Bras Cubas*.

Colonel Lágrimas forms part of this line of works that turns to non-chronological, non-systematic fragmentation to transcend local references and anthropological approaches, and incorporates the complexities of the imagination. The colonel isn't just an antiquarian, but an historian in a new and full sense. At the beginning of the twenty-first century, as we attempt to understand our past and present in an increasingly technological age, the portrayal of his confusions seems illuminating.

Here, we reach a final quandary. The more beautiful the telling, the more likely a story will be received as true. Truth and aesthetics, the question of whether a theory is accepted because it proves something real or because it has a certain allure, threaten to become synonyms throughout the project. Fact and fiction are intertwined, so that history becomes a myth, a cipher, a colourful geode. In the middle of the Amazon jungle in 1824, the colonel notes, Johann Moritz Rugendas picks a rock off the ground. In it he sees 'the strata of a past time, history turned pure geometry, and thinks: it was with this rock that Cain killed Abel'.

The distillation of various ingredients into a single potion of meaning and beauty may sound like magic, and there is a sort of alchemy at work here. Not only does Grothendieck think of his work as a charm or amulet, an enchantment, but he also literally turns to magical explanations in his efforts to understand history, realizing historiography is simply another kind of magic. His various works of reference include the *Spiegel der Kunst und Natur* (Mirror of Art and Nature), written by Stephan Michelspacher in 1614, and the *Natural System of Colours*, written by Mary Gartside in 1766, outlining a theory that 'construes the colour palette in a way that never ceases to seem magical'. This magic can mean a sense of wonder, a replacement of certainty with stardust, a questioning of linear chronology, a disintegration of time into rearrangeable quantum units.

Reading Fonseca, and writers like him, we find ourselves in a world that is not our familiar old sphere, but the y = 1/x equation of a hyperbola, as in Christopher Priest's novel *The Inverted World*. No simple narrative is offered to us. But the kaleidoscopic proliferation of visions investigates all the ways the intellect can get tied up in silk knots, while eluding the merely erudite. *Colonel Lágrimas* is a book you can read multiple times, and each time discover new meanings, new rays of interpretation radiating from the anecdotes. The seductive danger of such a tale is that the reader, too, is caught up in trying to assess and decipher the damage.

The Formula (after Carlos Fonseca)

Bivouacked in a cave on Pic Maudit, he tries to find the equation. He wants to find the expression that captures all of himself, the century, the landscape around. He sets about piling up stones, arranging them loosely but with purpose, paying close attention to the small gaps between them. Apocrypha is what the stones are, anecdotes whose meaning no one quite understands, lucidly told but lacking that indefinable thing that makes you say, yes, this one here, this is the formula. In truth, finding the formula isn't the hard thing. Recognizing it as the formula is what's difficult, since likely it will arrive in the form of metaphor. He writes potential formulas in black with felt-tip pen, paints a stone blue, writes a single line on the surface of another. *What if I'd been a father?* He scrapes a third all day with a piece of raw leather, the syncopated sound of stone on stone suggesting he's getting closer to something, ever closer — but no, more stones. On one, he writes: *What if I'd been a mother an animal a plant a wire a cable a different person*. On another, *Inconnue, résistance!* The piled-up stones accompany him as he putters about the ledge. He picks up his canteen. Moving toward the pebbles, he lets the canteen tip as an executioner would drop a hatchet, without emotion, love, hatred. The water falls over the rocks, and as it trickles down he wonders if it has to fall that way, if there's any freedom in that falling. Compelled by its own cascade, the water glazes the stones, clarity. The formula might appear if the water never stopped running, he thinks. If it kept trickling through for infinity, maybe all his meaningless scribbles would be washed away, so the hidden message in the stones could emerge. Something clean and simple, something that explained everything ...

Here and There

imagining utopias in language / Everywhere

Technology isn't a terminus, but a means. As such, it forces one to think about utopia: what those means are aiming *toward*. What utopias can be imagined?

The Utopia of Satire?

At the heart of satire is a stereotype, a simplistic dichotomy, an obvious truth or an unquestioned form ready to be taken to its limits and dismantled. The thirteen stories in *Insurrections,* the first book by Rion Amilcar Scott, satire editor of *Queen Mob's Teahouse* and Professor of English at Bowie State University, are set in the imaginary Maryland town of Cross River. This founding act of playfulness is the departure point for the stories that follow.

Most feature some version of the author himself, who temporarily assumes another gender, another age or another set of life experiences with the wondrous flexibility permitted by fiction. It's clear that Amilcar Scott feels deeply and that he has a deadpan sense of humor, capable of keeping him balanced on the shaky beam between intelligence and ridiculousness. For satire doesn't just mean 'being funny'; it's an existential mode that allows one to take on both joyful and painful subjects from inventive, oblique angles, allowing one to make almost anything one's subject with good humor, precision and grace.

In Amilcar Scott's stories the ordinary becomes legend, and whatever self-righteously thinks itself sacred is taken down a few notches. One might say that he picks up black tropes about bibles, barbershops and boxing matches in order to explode or comically deflate them, if this did not sound like something from a Sociology of Race Relations 101 class, and if there is one certainty, it is that this book is far more complex, far more *interesting,* than what an over-serious academic text might make of it. Skewering the excessively sanctimonious, the author describes the church rector, who 'wore a black turtleneck and jeans', and whose 'bald head appeared freshly moisturized'. Someone asks if Jesus was black, and the man becomes distinctly uncomfortable. 'I imagined Black Jesus up in Heaven laughing like hell,' the teenage narrator writes, about to go for his Confirmation but more worried about whether or not his crush, Alana, will be impressed by his pinstriped suit. Whether or not Jesus is in fact black, and there is no 'confirmation' of this, the narrator is sure he has a black sense of humor.

Despite the banter, a strong moral compass holds. Characters behave with courage, whether the stands they take are for the ages or pointless. 'I've never been one to watch weather reports. It's more honorable to take the weather as it comes,' insists the narrator in *Everyone Lives in a Flood Zone*. In *Good Times,* a Muslim dad named Rashid, feeling lost and seriously considering suicide, thinks of everything he could have done better. He fantasizes about dropping off his baby boy at daycare, dressed as the Cookie Monster: 'I'd be the most popular father who ever existed, showing up shaggy and blue with a tin full of snickerdoodles. That was the dream.' Often, plans don't come off quite as envisaged but, talking to his baby, Rashid explains that difficulty is part of life. 'Forever, huh? I was going to name Luce forever, or rather, Samad, one of the ninety-nine names of Allah — Al-Samad, the eternal. But then I started to think about eternity, what a curse if you're not God, right? My man God doesn't have holy rent and holy bills to pay.' In this tale, the inverse of the deflationary lampooning in other stories, daily life becomes its own feat, creating its own lore.

In 'The Legend of Ezekiel Marcus', complexity arrives in the form of the female, a simultaneous goddess and source of fear for the young narrator. If the greatest fear is fear of the unknown, then for the inexperienced boy telling the story, women are the supreme enigma. Battling doubt and shame, he is, along with his classmates, unsubtly appreciative of the curves of the female body. ('Kelli's breasts. What was it about them that caused such derangement?') Things take a dark turn. Kelli incurs the wrath of other girls who want to 'rip her weave out'; a well-liked art teacher is quietly removed by the school for commenting on Kelli's curves; upset by the seeming randomness and injustice of everything, one of the boys attacks the only white teacher. Leaving school, the boy disappears, possibly becoming a hero, possibly making millions as a soccer star, possibly dealing drugs, possibly sprawled on a beach in the Caribbean or Europe, hiding from the cops. What is real, what is myth? High-school humiliation here becomes the basis for legend.

Despite the heavy content, Amilcar Scott takes a real joy in language, and his book has a variation in its rhythmic flow that is delightful in its piquancy, possibly more than a purely elegant lyrical text would be. Sounds, voices and the oral tradition are important, and quotes are sampled from pop culture, song lyrics, slang and sacred texts. In the stories there are frequent changes in register, places that are like a boy's voice cracking and dancing off for a second into a higher key; deeply distressing at the time, hopefully hilarious in retrospect, an in-between moment signifying some sort of change in progress. The weird shuttling between high and low (*a dark-hued maze of haberdashery* coexists with *C is for Cookie, He who is without sin can cast the first stone* and *Come on and bounce those big things, baby*) just goes to show how language can be both a battlefield and a playground.

In this linguistic terrain, people often do not know quite how to talk, or are learning what words to use, because they are immigrants, or new to the neighborhood, or being educated into a new idiom. In 'Juba', the narrator is mistaken for the drug dealer of the title and beaten up. He decides to track down the real Juba and, to his surprise, finds him engaged in literary translation, trying to turn classic works into the street language spoken around him. The narrator asks if the book in front of him is a dictionary.

> Naw ... naw ... hell naw, this ain't no damn dictionary. The people ain't ready for that. For like twenty years, I been translating the Bible into Cross Riverian, as you bougie niggas like to say ... Y'all spend a lot of time translating from English to Cross Riverian and back in y'all heads. Y'all just don't know it. Niggas ain't slow, they just translating ... I'm gonna do the Koran next, and then the Bhagavad Gita. I already did the Heart Sutra. Did that shit to warm me up. I got a rack of other sutras to do, but that's a ways off ... Got folks mailing me new words in exchange for 'Dro. The police might catch up to me before I'm done.

Is this man a lost cause or some kind of erudite rebel prophet? Who knows, and maybe the distinction has little basis, anyway.

Party Animal: The Strange and Savage Case of a Once Erudite and Eloquent Young Man is a diverting fake case study of Reverse Animalism, the supposed phenomenon of intelligent young black men reverting to animal behavior. With unintentionally amusing foot-in-mouth self-consciousness, the authors describe the activities in question with a high degree of detail, such as visits to nightclubs in which velvet ropes are crossed 'in order to fondle and "freak dance" (i.e. rhythmically gyrate the male genital region against the buttocks or genital area of a partner in an erotically stimulatory manner)'. The authors of the text claim they will avoid black stereotypes like the happy naked, primitive and Hottentot Venus, but in referring to their omission so carefully and at such length, they show just how painfully present these ideas are in their minds. A postmodern pastiche, the erudite text cites non-erudite sources to support its argument, blooming with footnotes, caveats and complaints into a gaudy textual bouquet of dubiousness. The set form of the academic essay becomes ripe material for burlesque.

In Cross River, a diverse place, there are 'bougie nigs' who think of themselves as 'Cross Riverians'. There are Salvadoreans who think of themselves as 'Riverbabies'. And there are Muslims, often seen by neighborhood blacks as a desirable other. This becomes a source of self-criticism: are blacks doing to other groups what whites do to them, turning them into mysterious stereotypes? One moonstruck young man marvels at a woman: 'First, the dazzling eyes, two burning brown and green sparks dancing on her face. Then, the veil.' (In real life, Amilcar Scott is married to a Muslim woman, whom he acknowledges for loving him and 'critiquing every syllable'.) Whatever the case, all kinds of people come into contact while getting on with their daily lives. 'Black' comes to absorb all kinds of meanings: 'Our swarm, it move like a flock of birds. All these beautiful black people in motion. Moving and shifting with a kind of intelligence. When we reach the destination, we just know it.'

The last story, 'Three Insurrections', ties the collection together. Here, time is shown to be a material capable of stretch and dilation, in which family, national and universal histories merge. The narrator is the author's father, but he's far more than just one character. By chance, he finds a book in the library with flames on the cover, called *Three Insurrections*. It intrigues him so much that he misses his cricket match to read. Later he speaks:

> The Haitians have an insurrection. The Riverbabies — the Cross Riverians — they have an insurrection. And there is one to come and it's mentioned with the ones that happen like it's a piece of threaded gold passing through the garment. I don't see my name, but I see me. I see you and you don't even exist ...
>
> Something make me left that book in the library, though.

> Maybe it was too much to take, the way it make my mind spin and spin. I wish to hell I had grab it and run. From then on, Cross River is burn in my brain … But what kind of people is this? I think. These Cross River folks bloody they masters and live free like they not afraid. The book talk about the Haitians too, I hear about them plenty. The Cross River negroes is new to me. I see my island in a footnote. Some Cross Riverians set off through the Americas, trying to export insurrection. Some even settle in Trinidad, the beauty just hold them, even though they have slaves all over to free …
>
> Something about this book, Kin, you don't read it. You read it, but it make you live it, like a dream. I come a Haitian that day, and then I come a Cross Riverian. And just like a dream I live that third insurrection too, but when I close the book, when I leave the library, I forget what it's like in the third insurrection, and then I must spend the rest of my life chasing it down.

Amilcar Scott listens to his father talking, and thinks he has no stories as vivid as that. So he writes this book. The relationship between his memories, history, imaginary city and ideas of 'insurrection' is never completely clarified, and it's possible the author isn't clear himself. But these babbling literary mash-ups, sparkling with wit, a self-deprecating sense of humor, a keen sense of history and a range of voices, open new doors of possibility. On the cover, birds riot in a splendor of crimson and dark gray, flustered by something unseen. Perhaps, in language, a third insurrection has already begun.

Confronted with an immensely complex social system that it would be impossible or undesirable to take on directly, what should one do? Raised in a certain community but educated with university values, self-conscious about his literary position, Amilcar Scott faces the question of what stand to take, in fiction and reality; and if one should take one at all. If one speaks out, one will be crushed immediately. If one feigns ignorance to further an agenda within the system, using its own mechanisms, it is all too easy to succumb to cynicism. The creation of an imaginary city is a way to circumvent the excess gravity and political shrillness of real life, to come at situations from a more offbeat angle. Satire, in these stories, appears as a third way: a means of creating a fictional life for oneself that is whimsical yet self-interrogating, sustaining argument but with soul breathed into it through humor and a healthy dose of silliness.

The Utopia of Sincerity?

Los Angeles is a city of sprawl and sunshine, but it can also be a very lonely place. When Maggie Nelson moved out there from New York to teach at CalArts, heartbroken, she wrote her collection of prose poems *Bluets*, about her love affair with the colour blue. All her personal experience somehow became abstract in that locus of freeways, museums and well-watered lawns. Janice Lee's fragmented 'essays', inspired by Nelson and other past teachers like Eileen Myles, along with French theorists such as Gaston Bachelard, her dead mother and musings while visiting favorite places along the coast, begin from a similar place of loneliness. This is a book about sadness, about the ways in which a young woman understands and dissects her emotions, and transforms them into art.

Several of the pieces were previously published on Lee's online magazine *Entropy*, and the independent press background is clear in the work's formal innovations. Not all of the sentences or phrases work completely, or at times seem to express 'obvious' sentiments. Yet this doesn't seem to matter, as it's the accumulated effect of phrases that's of value here, not any individual quote. Lee seems to think of the space of the blank page as a kind of house, one in which the placement of furniture (the words) is only important to a certain extent, as words, like furniture, can be rearranged. And so, while she insists memories must be localized, as in her meditations on the Salton Sea, there's a constant dislocation within the texts, possibly intentional. Adjective often doesn't quite sync with noun; there's a sense of searching and incompleteness, quotation and repetition, a casting about for the right word; and all this is part of the point. A wash of words breaking against shore, searching over and over again to be 'the' phrase, never closing off meaning or declaring itself the end, receding and returning, eternally capable of shifting form into something else. Empathy and heartbreak consume and recreate themselves; sadness feeds on sadness, blue on blue.

Here argument is poetry, and emotion becomes process. Lee dismantles language into almost mechanical parts, thought diagramming into other thoughts, and this abstraction makes it an art text. In an essay about preparing a reading for Los Angeles County Museum of Art (LACMA), Lee worries about this way of thinking of her body and emotion as elaborately structured pretexts for action, and the attraction to structure as the inverse of sprawl, drift and disconnectedness is a recurring theme. In *Backpacking, Point Reyes, Driving,* Lee says the sestina, in which lines end in patterned repetition, is an appropriate poetic form for the childhood place where she finds herself, a place that 'tends to conjure and reconjure ghosts, the uncanny repetition that induces haunting, déjà vu, strange warpings and relocations of memory ... Echolocation.'

In the same essay Lee sees a flock of quail, 'unafraid, beckoning or mocking or completely apathetic to whatever it is I am doing'. This natural world exists beyond her personal concerns; the human and her anguish are irrelevant to the non-human world. Los Angeles, the financial system and the Anthropocene reality in which we find ourselves, do not care about our emotions; abstract intelligence is what is most highly valued. Yet, throughout the book, Lee expresses her nostalgia for a way of life that is animal, pure non-thinking. She believes that the poetic stance of openness to one's surroundings, and the intensity with which one perceives and expresses consciousness, can transform a tree into a haiku, a peacock into a pantoum.

This processing of emotion into poetic prose is a tremendously valuable project. It is also a *trompe l'œil,* as this insistence on empathy becomes so divorced from any specific person that it, too, becomes cerebral, an object. The temptation to mysticism, and desire to simply embrace nature, are framed in the most abstract, almost aphoristic prose. That love could be for anyone. That sky could be anywhere. Triggers for memories could come from direct life, someone else's life or stories and myths. Lee herself admits this. 'Even in this state I can recognize the shameful sincerity of my situation. I do not miss you or even you. I miss the sensation of you. I miss the warmth, the questions,' she says in *Mornings in Bed*. Her way of writing is a way of seeing the world, an embrace of the ahistorical, unattached life at the western edge of a young country. 'There are no ancient tribal feuds, no wounds, no blood. It is less absolute, perhaps. But better' than the history-obsessed Old World, as Chris Kraus claims in the last lines of *Torpor*. 'The sky isn't blue', Lee's rallying cry of possibility inhabits the same mental space. For Lee, the sky can be (should be?) different every time.

In Los Angeles, where the sky is always blue, this is a radical denial, but its 'anything goes' ethos is also the most LA statement there is. The city is central to Lee's fragmented accounts, as it is an urban nexus, a place where one can wake up beside someone and never see them again; where relationships require advance planning; where all is atmosphere, gridlock, freeways, immensity, movement; where life is different for everyone, even different for oneself from one day to the next. LA is an unquotable haze, just as in this book, almost no sentence can be quoted, but quoting isn't what it's about, even if Lee herself occasionally hat-tips the authors she likes, such as Kenneth Patchen, Jaime Saenz and Pierre-Jean Jouve. What it's about is an attitude, the creation of an atmosphere. This attempt to articulate emotion in a unique way can reduce itself to trivial thoughts, which provoke despair precisely when one realizes *nothing new can be said*. The lyre's been strummed since the beginning of time, the sky is blue.

Yet when Lee dares to flout the obvious, and say once again that she is sad, that the sky is not blue, something happens; this becomes an intrepid statement in itself. The imagination is rearranged; the insignificant becomes significant. As Lee writes:

> ... that is what love is, perhaps, a complete rearranging of the imagination, a complete infiltration of a subjectivity that seems to defer how images correlate with each other. Suddenly, what matters is the color of the sky. The direction of the stars. The speed of light. Significance and insignificance change places.

Perhaps that is what the inverse of heartbreak involves, too. In the utopian dystopia of Los Angeles, where the certain and uncertain, the vastness of the city and tininess of one's bedroom, the permanent and transitory, the inane and artistic, become irrecoverably confused, dichotomies and structures begin to lose meaning. Lee's sincere tributes to tears, rain and the places that mean something to her begin to refract infinitely in a mirrored hall of self-consciousness — perhaps one way, in a lonely city, to cope.

The Utopia of Lightness?

The first and most important mental habit that people develop when they learn how to write computer programs is to generalize, generalize, generalize. To make their code as modular and flexible as possible, breaking large problems down into small subroutines that can be used over and over again in different contexts. Consequently, the development of operating systems, despite being technically unnecessary, was inevitable. Because at its heart, an operating system is nothing more than a library containing the most commonly used code, written once (and hopefully written well) and then made available to every coder who needs it.

'In the beginning ... was the command line', Neal Stephenson

When *The Missing Slate* asked me to prepare a dossier for its poetry section, I chose to focus on poets from Bolivia, understood loosely to mean writers with some connection to the country. Why Bolivian? 'Bolivian writing' is a phrase that, like French or Mongolian writing, in itself means nothing. Poets work from a personal vision, or maybe with affection for a certain city. Any fervent propaganda on behalf of a nation is already some other project, like politics. Or is it? What is poetry, after all? Is it a text with rhythmic and aesthetic qualities? Is it life beyond the text? Is it, *Viz* pornography, 'you know it when you see it'? I'd rather avoid the snapping intellectual trap of defining terms, and use the time to read the poems themselves.

The real reason I've chosen to focus on Bolivian poetry is its convenience, an arbitrary selection method to read a wedge of the world's infinity of poems in greater depth. Several poets from the region struck me as compelling, and I wanted to know what I'd find going deeper. Poems from the past, poems by contemporaries ... I'd recently visited the country and was interested in all the stimuli that met my eyes, but what mattered now were the texts.

A dizzy series of days followed, reading poems on paper and Internet, and two impressions resulted. The first: there have been many paths abandoned. Individuals who experimented with the way they expressed visions, feelings or language, whose work was for some reason left by the wayside, not taken up. The second:

the poetry that struck the strongest chord in me started with a first-person voice and some element of reality like an object, before proceeding to personal reflections or fantasies based in imagination. The visions in different poems both overlapped and diverged. I began to think of all the words as proceeding from some grand archive, far larger than the mini dossier I've prepared here, packed with different ways of writing; either that or a well-stocked cellar with different intoxicating substances available for combination.

As a literary form, poetry is particularly free in its structure, rapidity and ability to incorporate novelties. Its texts are written in some context and may be linked into a 'social history'. But they can also be unloosed from this, understood in terms of a history internal to themselves or unique to the textual experience of the reader. The past is always a click away, full of texts and styles to be discovered, not in linear progression but with creative anarchy. Poems can be read 'out of context' productively, taken up and reappropriated in new context. This is what I had in mind while compiling these poems from the last century of Bolivian writing, translated into English.

Developmentalist Distractions

In 1959, literary critic Fernando Diez Medina called for a literature representing the national geography, which he divided into three regions: *valle, sierra, altiplano*. Literature that represented Bolivia as a country in potential, capable of industry yet nourished spiritually by a mysticism linked to its kolla and aymara past. 'The dawnlike awakening of the Bolivian people will find its support in aesthetic and literary renaissance,' he wrote.

At about the same time, Bolivian cultural minister Roberto Prudencio Romecín took a different line, praising the cholos, contemporary indigenous people who had migrated to the city. Yet he shared Diez Medina's developmentalist vision of literature. The collection *On books and authors*, which compiles essays written throughout his life, gives a good sample of his perspective. 'Very few have dedicated themselves to the history of our literature,' he begins one piece. 'Our new generations do not seem very inclined to the cult of letters,' he starts another. 'Rare are those among us who have dedicated themselves to pure philosophical speculation,' begins a third. 'The novelistic genre has not had many, or very good, cultivators among us,' he starts a fourth.

One could go on, but the point is clear. From this absence, a new generation of nationalist writers was expected to emerge. It was a call to arms, and the collected work is a dizzying attempt to alchemize gold from nonexistent materials. Can a country's literature really be built deliberately, like a railway forged of purely national steel? These critics, writing shortly after the 1952 revolution and subsequent reforms, were convinced it could. Even now, literary critics trace analogies between poetic movements and historical events such as the Chaco War, Movimiento Nacionalista Revolucionario, rule of the military junta, series of coups, 1982 democratic elections, hyperinflation, quadruple election of Víctor Paz Estenssoro, protests over the privatization of Sánchez de Lozada, disputes over natural gas reserves in the south, 2005 victory of Evo Morales, and focus on nationalization and indigenous rights. But recurring to lines of narrative history to 'explain' a style often has little to do with the way actual poets write.

Take the symbolist Jaime Saenz. He made pictures too, and they're a good place to start. In his skull drawings, a line can mutate to become a spiral, a nose, a snail, a laugh with rictus, a minimalist signature in the form of his initials 'JS' (which in one of those simultaneously prosaic and profound revelations that occur while reading poetry, this author realized are also her own). The assemblages bear titles as cryptic as they are obvious, once labeled. Skull of a man, skull of a lady, skull of a globe, skull in love.

Names applied after the hand has done its work, not before. Saenz took the same approach to his poems. A prolific writer, he never stopped filling notebooks between cigarettes and nocturnal meetings of his Krupp workshops in La Paz, where editing of literary magazines, musical performances, readings, black magic, or card games might take place. He didn't lay out a program before writing; the act of covering pages with his scrawl preceded any deciphering. Write first and theorize later, was the poet's manifesto; an idea that runs counter to 'developmentalist' ideas of literature that start from the void and consider creative work as indicative of national progress.

A romantic construction of the self, Saenz's attitude largely ignored the nationalist agenda. It wasn't that Saenz was unaware of official ideology. Just like the other two poets who, according to poet and academic Mónica Velásquez Guzmán, have marked the path for Bolivian poetry — Óscar Cerruto and Edmundo Camargo — Saenz held down a government job. Yet his poetic work deliberately adopted a mystic approach, in which the self is inscribed in the universe. He examined his own skull and interrogated his own mind to talk about the world. This gives his work a remarkable coherence. Although he wrote from the 1950s to the 1980s, his rambling style drawing on specific images — night, cold, comets, magnets, stones — as a means for expressing his inner vision remains consistent.

The 'non-political' nature of symbolist poetry allowed writers like Saenz to speak without reference to specific politics or people, and gave his work a sense of timelessness beyond context. This is likely why Diez Medina, Romecín and other critics tended to ignore it, in favor of focusing on the novel, a more effective vehicle for their preferred indigenous social realism. Certain symbols might be interpreted, faithful to authorial intention or not, as commenting on the politics of the moment, but it is also possible to read the poems with pleasure beyond these temporal indicators. If they are not obvious metaphors, symbols can permit the freedom of a compression of meaning. Most of the poems included here, by a variety of poets writing at different moments in Bolivian history, include these symbolic elements, and none are explicitly 'political'.

In the same way, one can read and analyze non-politically the imaginary jungles of Raúl Otero Reiche, the enigmatic landscapes of Blanca Wiethüchter, the 'parodies, inventions and blasphemies' of Humberto Quino, the death-obsessed bodies of Edmundo Camargo. These poetic themes not only possess a historical link to the moment at which they were written, but can also be cut and pasted from their historical moment into the present, as easily as 'command+C, command+V' can carry a Spanish text online to a Word document on my desktop for translation.

The Pop Turn

And what about the poetry written now, in the Estado Plurinacional de Bolivia of President Evo Morales? Despite nationalist measures, Bolivia is highly open to international influence. Its writers increasingly form part of an international network of connections and this, almost inevitably, brings the influence of pop. Poets (and short story writers, and novelists) want to write about Bolivia without writing about Bolivia, to pay attention to its local characteristics without stopping there. Rather than turning into solemn mystics like Saenz, they turn to irony. Contemporary poets often write with a wink with regard to their own position as poets. Pop is the self-conscious moment the object becomes aware of itself, when the raw material of the 'popular' becomes material to be worked and transformed, aware of its place within the industry, economic structure, system of scholarships, and network of writers and readers. Given this situation, what is the best tone to adopt?

Some contemporary poets, like Emma Villazón, have chosen a lyrical tone bordering the surreal, permitting reference to the real with a layer of mediation. A bird may carry the poet away in its belly, but some part of her country will always remain with her. (Villazón, whose work has a kind of dreamlike or mystic vibe, died returning from a book fair in La Paz, at the age of 32.) Other poets, like Julio Barriga, adopt the Saenzian aesthetic of the alcoholic vagabond drawn to the abyss, but do so with self-conscious pathos. Barriga goes about the bars of La Paz distributing mimeographed, hand-stapled copies of his poems, and talks openly of his love for Amy Winehouse. The self-consciousness and ironic humor in his tone are clear. This attitude, increasingly common, is a way of going on writing when excess solemnity seems incapable of the toleration and quick flexibility required in a society (and world) in flux.

A more frequent alternative to the 'pop' approach than Saenzian solemnity is in fact silence, the position of the well-read genius sure enough has already been said, that there is no further use putting pen to paper. In the 1930s, the brilliant and eccentric modernist Hilda Mundy — discussed in a previous essay — stopped writing after completing only one book, *Pirotecnia,* convinced sufficient literature existed and that the true avant-garde consisted of living her life. Some may find their ecstasy in this approach, but for those who want to go on producing, a different approach is required. In the essays that accompany the poems here, *júbilo* (joy) is the word novelist Edmundo Paz-Soldán uses to describe the work of Hilda Mundy; it is also the word Liliana Colanzi quotes in a poem by Julio Barriga. But between the two poets there is a world of difference.

Faced with the abyss, one can choose to be not silent but conversational. The deliberately kitsch visual poetry of Paola Senseve, the linguistic whirlwind of Sergio Gareca mixing references to Michael Jackson and Quechua poet Juan

Wallparrimachi; these are poets interested first of all in play, not international poetry festivals. Yet, they're aware those festivals are waiting. Bolivia's writers have one foot in, one foot out of international markets, and their work has a certain self-consciousness about it. Even if they do not approach their work as Bolivians, or as part of an explicit national project, they are aware of structures far larger than themselves (life, death, capitalism). They live in a place where state and culture are intimately intertwined, yet are tapped into foreign influences.

It's worth noting that an increasing number of Bolivian writers come from the middle classes and suburbs, which encourage minor idiosyncrasies and a love for pop, as well as a respect for the avant-garde, so easily co-opted by global markets. The suburb is a place that is anti-nostalgic, or that produces a generic nostalgia given to categories rather than specifics; the same categories that appear in identity politics, scholarship applications, and other middle-class institutions that encourage you to represent yourself as larger than you are. 'Be an individual with quirky characteristics that set you apart,' these institutions seem to say, 'but make sure you can also be located within a greater whole.'

Does the kind of poetry that results form part of the pop-focused, anti-magical realism 'McOndo' movement theorized elsewhere by Paz Soldán, himself a novelist? Has poetry, too, become 'McLiterature'? The term is playful; it does not have pejorative connotations. Like the artistic term 'Fauve', it embraces the elements once used to attack it. Perhaps this particular kind of self-consciousness is to some extent present in every developing country with a growing middle class, from Brazil to India.

Perhaps self-awareness even plays the same role for contemporary poets that alcohol did for Saenz; a distorting intermediary necessary to see the world 'better'. The lens of humor produces an altered state, and any decadent paeans on behalf of intensity and against indifference, attracted to blood and torture, focused on death, the body, visions, flames and oblivion, must be taken with a grain of salt. Contemporary poets are still interested in absence and presence, loss and love, but these elements are mediated. I read the irony in contemporary Bolivian writing as another variation of symbolist poetry; a tone suggesting that beyond the surface of what is said, there is always something more, an invisible metal hidden beneath the earth.

Toward Lightness

The 'pop' tradition, capable of making clear and even loving reference to Reebok, Faber-Castell or Twitter is an understandable reaction to sluggish chronicles saturated with local colour, eyewitness accounts and references to purely national tradition incapable of comprehending the bewildering and complex world of data in which young people now live, above all through the Internet. Pop is here to stay, and probably unavoidable. (One can choose to practice it responsibly.) But there's another way of connecting as well, not with explicit references but with sensibility. An awareness of what's being written in other places in the world and what was written in the past, in one's own country and elsewhere. A connection to the grand archive of the past that trickles into one's own writing. Both 'horizontal' and 'vertical' connectedness exist; the former entails a connection to contemporary market phenomena, the latter to international historical and literary texts.

In this sense, more indicative of the current direction of regional writing than the mystical verses of Saenz may be the 'light' poetry of Pedro Shimose, a Japanese immigrant with international connections to Spain, capable of moving vertiginously from the archival trawling of (invented?) Machiavellian papers to the small pleasures of drinking coca tea. This sense of freedom, the ability to include anything one likes from colloquial speech to Florentine parody, is conducive to participating in the grand archive. Freedom in not just citation but attitude; the poetry I like best is full of good-humored openness, where nothing is prohibited, and the world is a beloved secondhand bookshop through which we are free to wander.

A few months ago, over a beer with some writers in Santa Cruz de la Sierra, a conversation that flitted from Argentine elections to riots in Sucre, I remember thinking: we're drinking Paceña, but it could well be Quilmes, Corona or Jupiter. We're talking about the Pope, but could well be discussing Li Yu or Anna Karenina.

Bolivia is an increasingly prosperous country with a growing middle class, widespread Internet connection even in the tiniest pueblos, and a population of educated and mobile young people with academic scholarships and international travel experiences. Traditional geographical and ethnic distinctions have begun to blur. A writer may fill page after page alone in her room, then take that notebook to a bar for a reading, one she will perhaps repeat later on in New York, Santiago, or Moscow. Perhaps — is this just a fantasy? — the poets of Bolivia form one small part of a worldwide movement in which nations as we know them disappear, along with progressive 'developmentalist' thinking, to leave only the pure flow of cash, art and ideas.

The Utopia of Affection?

Recently, I've been thinking about the idea of a state of eternal bliss that persists. At the Pierre Bonnard exhibition called 'Arcadia' at the Legion of Honor in San Francisco, the images I like best are not necessarily the official representations of paradise, elaborate works displayed in the dining room of the patrons, a millionaire couple. I prefer the painting 'Le Cannet', a view from above of rooftops and trees. In it, Bonnard's colours are extremely intense, his shapes precise. Although no people can be seen, it's easy to imagine them in their homes taking baths, stroking cats or spending time together, just as they do in Bonnard's other images.

The rolling hills of Le Cannet are not so different from those of California, and I'm not sure whether to call this beauty 'wild and tangled' or 'tame'. I suppose in the end it's a question of comparison, what reference wild and tame are being compared against. Just as in Goethe's colour theory, perceptual experience is what ultimately matters.

Colours ... everything is in bloom here, at the start of summer. My mother likes to garden and knows about flowers. When we leave, I ask her what the one in front of the museum is called, a plant whose base looks like an artichoke with a purple bloom emerging from it. For some reason it's very important to know the names of things today. Hen and chick, she hazards. The big flower is the hen, the nested leaves surrounding it below are her children. Then she tells me more about the duplex she's renovating in San José, which she considers her art.

In ancient Greek 'arcadia' referred to a peaceful wilderness, a place of natural splendor and harmony. Here at home for a while, recuperating after illness, I feel this is a close substitute on our planet. My mother has a garden with flowers in many colours, and there are rows of lemon trees. I like to walk past and touch their leaves and breathe in their scent. Unlike the events of other months, they seem something real. If I scatter birdseed I can watch the ringtail doves bob their heads and peck the earth. They are calm here, unlike the pigeons in the plazas of Argentina or Bolivia. But I suppose the distinction has less to do with nations than with country vs. city animals. I imagine the birds of San Francisco or New York City are vicious, too.

For some Marxists we exist only in the minds of other people, and individual mental existence and works have no meaning. Perhaps there is some incomplete truth to this. Yet, somehow I can't be convinced by it, or need to believe in some possibility that is stranger and more personal. The convolutions of time and memory seem far more fantastic than any of us realize. Perhaps someday we might even be recreated based on the works we leave, so our writing ensures survival in a literal sense. I wonder if there is a way to transform the stories written by my grandmother, for instance, into something that can survive her once she is gone, something with a sustainable life of its own.

One story is about a painting in her house of a brown dog with paws pressed tight together and a collar around its neck. It says R O S E in big letters, an enigmatic name. It's allegedly by a 'primitive' painter in New York who, as Grandma (always politically incorrect) pointed out in a whisper, is black. But famous, she would have me know. The man gave her that painting long ago, before she met her husband. Is she the rose? Or if not, who is? All the details must be in her journals ... Her other story, to my surprise, is about me. I live in Argentina and am married to a cattle baron who is incredibly wealthy. Supposedly, he is in command of a troop of men who spend their days chasing down beasts rodeo style, while the baron has all the time in the world to embrace me.

These are her only two memories left. I know one is not true, and probably both are confabulations. But they are who she is now and for her they are real, constituting her consciousness. If I could bring them into reality, perhaps the dog and cattle baron might imagine her, so she continues to exist even after her body is gone. Is this crazy? Can an imaginary imagine a reality? 'You turn to literature as a consolation instead of tackling problems head on,' my mother always rebukes me. This is probably true, but it changes nothing, does nothing to allay my confusions or answer my questions.

At the entrance to the road where my family lives, someone has been moving the stones every day, positioning them in different arrangements. I wonder if the activity is being documented, and if to be considered art it requires a register as photograph or film. If it were put in a gallery or had pretensions to enter art history, it would be required. In this part of the world, where utopia is technological and progress quantifiable, the idea of activity for its own sake seems almost subversive. But maybe Arcadia means precisely this, doing things for the sheer pleasure of them with no need for record, no egoism. Along with Le Cannet, another of my favorite Bonnard images is a white cat whose back arches so high it seems not to have a neck. Its eyes are squeezed shut and its tail is a squiggle, pure joy without an ulterior motive.

For me, a real Arcadia would involve complete trust and be something like a cathedral of belonging. The space first appeared to me in visions and dreams. When I started thinking about it consciously I wondered if the loss of creative tension would also mean the end of production. Perhaps art must emerge from an emotion like despair, nostalgia, rage, resentment or fear, or the recollection of one ... But this seems a silly, artificial worry. I think even in Arcadia we'll continue to make things. Maybe not installation art featuring violence, depictions of torture or bodily illnesses, but small intimate scenes pulsing with colour, celebrating affection. The idea of this eternity haunts me, time without beginning or end, in which no one ever dies and love survives forever.

Epilogue

Those Hills Behind

In sixteenth-century woodcuts and paintings you see them often: those hills behind. Starting off craggy, they descend to rolling. Maybe a zigzag river winds its way through, along with a hamlet tucked away or a castle going about its business, unaware of events in the foreground: a saint dying, a battle, a visitation, an important session of stargazing. Often, especially in times of uncertainty, I've thought of those hills behind. They hold some magic, a promise of peace and joyful existence far from grand events and ostentatious ceremonies. Do the hills exist simply to fill space, or do they serve a purpose of some kind? If they weren't there, our perception would be entirely horizontal, but the addition of those hills gives the picture a sudden vertiginousness. In a religious woodcut invoking the heavens, this seems immediately appropriate, but even in non-religious context the hills add something gothic, a promise of drama; one that isn't immediate but time-lapsed, in an odd, unpredictably staggered way.

All my life I've collected images of those hills behind, out of simple curiosity rather than any particular purpose, though the thought has occurred to me that were one to plot a literary crime, those hills behind would be ideal bearers of a message. Rivers in the background of a woodcut series might carry the plan for political insurrection, or spell out an occult theory. They might suggest concealed meanings or consequences about what's happening to us closer in the frame, a sneak peek at some time before or after the present we're meant to see, contained in the same physical space.

What do those hills behind have to do with science and technology? In a sense, everything. While writing the essays compiled in this book, loosely gathered as 'imaginative responses', the image of those hills behind hovered as guiding metaphor. At a confusing time in which people take Ballard, Baudrillard and *Black Mirror* as their gods, perhaps there are ways that foreground can precede background, subject precede structure, life precede concept, utopia precede the idea of utopia, science precede magic, technology precede theology. Perhaps Walter Benjamin's messianism can still suggest rich ways of understanding in this present, in which 'theological' (or 'magical' or 'prescientific') and 'scientific' ideas are capable of interacting in ways that appear contradictory or unexpected going by a strict 'make X to get Y' mentality.

Can non-scientific modes of thought be a result of science rather than precursor? Can they be responses rather than prefigurings? Can 'those hills behind' create the events foregrounded? Is it possible for an effect to precede its cause? Writing about a novel in *Infinite Fictions*, David Winters says that 'echoes and ghosts predate crafted characters, they can't be pinned down to one point of origin; they emerge from what the work opens onto.' For him, literature can fold away from the words spoken, to interact with what is unsaid, drawing back from an event.

One need not state directly, but write about and around, in a searching and multilayered fashion. Thinking in lateral, non-positivist, indirect ways, one can begin to engage with the ghosts of an occasion, starting with its imagined resonances and effects. Even as one enjoys the present, one can remain attuned to traces and echoes, histories and premonitions.

Those hills behind show up very strikingly in the 1958 print *Belvedere*. Since this is Escher, there are all kinds of strange, contradictory and impossible angles at work, complicating and enriching the picture with foldings upon foldings. A Renaissance-clad, floppy-capped man on Level One stares at the distance into those hills beyond; a solemn woman in a long dress on Level Two does the same, but at a different angle, the object of her gaze elsewhere.

What can she be looking at with such joy? Could 'those hills behind' exist behind us too? Surely the winding river in Escher's image continues up and circles into a landscape we cannot see. Who is looking in our direction, how can one attempt to find out? Are technical approaches useful, or are there other ways of knowing? Can one envision a near-present that isn't a dystopia? The brain begins to craft theories evermore baroque — one is tempted to simply plunge into the hills — but to do so, it's first necessary to turn around. What will we see there? Will a bird with never-before-glimpsed plumage fly toward us, blown our way from the future, cawing prophecies of events to come?

Impossible questions to answer right now, perhaps, but it's clear that whatever is going to occur will require new approaches. You are afraid to look, I know. So am I. Come, shut this book and let us go explore the territory.

Zero Books

CULTURE, SOCIETY & POLITICS

Contemporary culture has eliminated the concept and public figure of the intellectual. A cretinous anti-intellectualism presides, cheer-led by hacks in the pay of multinational corporations who reassure their bored readers that there is no need to rouse themselves from their stupor. Zer0 Books knows that another kind of discourse – intellectual without being academic, popular without being populist – is not only possible: it is already flourishing. Zer0 is convinced that in the unthinking, blandly consensual culture in which we live, critical and engaged theoretical reflection is more important than ever before.

If you have enjoyed this book, why not tell other readers by posting a review on your preferred book site.

Recent bestsellers from Zero Books are:

In the Dust of This Planet
Horror of Philosophy vol. 1
Eugene Thacker
In the first of a series of three books on the Horror of Philosophy, *In the Dust of This Planet* offers the genre of horror as a way of thinking about the unthinkable.
Paperback: 978-1-84694-676-9 ebook: 978-1-78099-010-1

Capitalist Realism
Is there no alternative?
Mark Fisher
An analysis of the ways in which capitalism has presented itself as the only realistic political-economic system.
Paperback: 978-1-84694-317-1 ebook: 978-1-78099-734-6

Rebel Rebel
Chris O'Leary
David Bowie: every single song. Everything you want to know, everything you didn't know.
Paperback: 978-1-78099-244-0 ebook: 978-1-78099-713-1

Cartographies of the Absolute
Alberto Toscano, Jeff Kinkle
An aesthetics of the economy for the twenty-first century.
Paperback: 978-1-78099-275-4 ebook: 978-1-78279-973-3

Malign Velocities
Accelerationism and Capitalism
Benjamin Noys
Long listed for the Bread and Roses Prize 2015, *Malign Velocities* argues against the need for speed, tracking acceleration as the symptom of the ongoing crises of capitalism.
Paperback: 978-1-78279-300-7 ebook: 978-1-78279-299-4

Meat Market
Female flesh under Capitalism
Laurie Penny
A feminist dissection of women's bodies as the fleshy fulcrum of capitalist cannibalism, whereby women are both consumers and consumed.
Paperback: 978-1-84694-521-2 ebook: 978-1-84694-782-7

Poor but Sexy
Culture Clashes in Europe East and West
Agata Pyzik
How the East stayed East and the West stayed West.
Paperback: 978-1-78099-394-2 ebook: 978-1-78099-395-9

Romeo and Juliet in Palestine
Teaching Under Occupation
Tom Sperlinger
Life in the West Bank, the nature of pedagogy and the role of a university under occupation.
Paperback: 978-1-78279-637-4 ebook: 978-1-78279-636-7

Sweetening the Pill

or How we Got Hooked on Hormonal Birth Control

Holly Grigg-Spall

Has contraception liberated or oppressed women? *Sweetening the Pill* breaks the silence on the dark side of hormonal contraception.

Paperback: 978-1-78099-607-3 ebook: 978-1-78099-608-0

Why Are We The Good Guys?

Reclaiming your Mind from the Delusions of Propaganda

David Cromwell

A provocative challenge to the standard ideology that Western power is a benevolent force in the world.

Paperback: 978-1-78099-365-2 ebook: 978-1-78099-366-9

Readers of ebooks can buy or view any of these bestsellers by clicking on the live link in the title. Most titles are published in paperback and as an ebook. Paperbacks are available in traditional bookshops. Both print and ebook formats are available online.

Find more titles and sign up to our readers' newsletter at http://www.johnhuntpublishing.com/culture-and-politics

Follow us on Facebook at https://www.facebook.com/ZeroBooks

and Twitter at https://twitter.com/Zer0Books